The Complete Guide to Building Your Own Home

and Saving Thousands on Your New House

By Corie Richter

THE COMPLETE GUIDE TO BUILDING YOUR OWN HOME AND SAVING THOUSANDS ON YOUR NEW HOUSE

Copyright © 2009 Atlantic Publishing Group, Inc.
1405 SW 6th Avenue • Ocala, Florida 34471 • Phone 800-814-1132 • Fax 352-622-1875
Web site: www.atlantic-pub.com • E-mail: sales@atlantic-pub.com
SAN Number: 268-1250

Library of Congress Cataloging-in-Publication Data

Richter, Corie, 1947-
 The complete guide to building your own home and saving thousands on your new house / author, Corie Richter.
    p. cm.
 Includes bibliographical references and index.
 ISBN-13: 978-1-60138-243-6 (alk. paper)
 ISBN-10: 1-60138-243-X (alk. paper)
 1. House construction--Amateurs' manuals. 2. Building--Superintendence--Amateurs' manuals. 3. Contractors--Selection and appointment--Amateurs' manuals. I. Title.
 TH4815.R53 2009
 690'.8--dc22

                          2009032343

Printed in the United States

PROJECT MANAGER: Melissa Peterson • mpeterson@atlantic-pub.com
COVER AND JACKET DESIGN: Holly Marie Gibbs • hgibbs@atlantic-pub.com
INTERIOR DESIGN: Samantha Martin • smartin@atlantic-pub.com
ASSISTANT EDITOR: Angela Pham • apham@atlantic-pub.com

Printed on Recycled Paper

We recently lost our beloved pet "Bear," who was not only our best and dearest friend but also the "Vice President of Sunshine" here at Atlantic Publishing. He did not receive a salary but worked tirelessly 24 hours a day to please his parents. Bear was a rescue dog that turned around and showered myself, my wife, Sherri, his grandparents Jean, Bob, and Nancy, and every person and animal he met (maybe not rabbits) with friendship and love. He made a lot of people smile every day.

We wanted you to know that a portion of the profits of this book will be donated to The Humane Society of the United States. *–Douglas & Sherri Brown*

The human-animal bond is as old as human history. We cherish our animal companions for their unconditional affection and acceptance. We feel a thrill when we glimpse wild creatures in their natural habitat or in our own backyard.

Unfortunately, the human-animal bond has at times been weakened. Humans have exploited some animal species to the point of extinction.

The Humane Society of the United States makes a difference in the lives of animals here at home and worldwide. The HSUS is dedicated to creating a world where our relationship with animals is guided by compassion. We seek a truly humane society in which animals are respected for their intrinsic value, and where the human-animal bond is strong.

Want to help animals? We have plenty of suggestions. Adopt a pet from a local shelter, join The Humane Society and be a part of our work to help companion animals and wildlife. You will be funding our educational, legislative, investigative and outreach projects in the U.S. and across the globe.

Or perhaps you'd like to make a memorial donation in honor of a pet, friend or relative? You can through our Kindred Spirits program. And if you'd like to contribute in a more structured way, our Planned Giving Office has suggestions about estate planning, annuities, and even gifts of stock that avoid capital gains taxes.

Maybe you have land that you would like to preserve as a lasting habitat for wildlife. Our Wildlife Land Trust can help you. Perhaps the land you want to share is a backyard— that's enough. Our Urban Wildlife Sanctuary Program will show you how to create a habitat for your wild neighbors.

So you see, it's easy to help animals. And The HSUS is here to help.

THE HUMANE SOCIETY OF THE UNITED STATES.

2100 L Street NW • Washington, DC 20037 • 202-452-1100
www.hsus.org

DEDICATION

"This book is dedicated to the people of New Mexico who patiently taught me so many of the old and new artisan skills, including the craft of adobe and green building. Most of all, I learned to love the people, ethnic diversity, slow pace, and realities of life."

TABLE OF CONTENTS

Introduction 13

Chapter 1: Getting Started 17

1.1: Why Build Yourself? 17

1.2: The Benefits of Building Yourself 21

1.3: Misconceptions About Building Yourself 25

1.4: Your Role as the General Contractor 32

1.5: Finding and Managing Subcontractors 36

1.6: Going Green and Environmental Considerations 39

Chapter 2: Preparing for the Process 43

2.1: Planning and Organizing 43

2.2: Budgeting Your Project...46

2.3: Hiring a Contractor 48

2.4: Dealing with Insurance Issues.......................................52

Chapter 3: Planning and Scheduling 57

3.1: Planning a Timeline 57

3.2: What Types of Schedules Are There? 60

3.3: Why Should You Plan Accordingly? 63

3.4: Act As If You Are a Business 65

3.5: Planning Steps...68

3.6: Essential Tools and Equipment....................................70

Chapter 4: Financing Your Project 75

4.1: Types of Construction Loans...75

4.2: How to Use Debt to Your Advantage.............................78

4.3: Shop for the Right Terms 79

Chapter 5: Finding a Building Lot 83

5.1: Location and Geography...83

5.2: Finding the Perfect Building Lot 86

5.3: What is the Difference Between a Lot and Land? 91

5.4: What is the Value of Your Building Site? 92

5.5: Should You Deal With Tearing Down Property? 95

5.6: Buying Your Building Site 98

Chapter 6: House Plans   105

6.1: Designing Your Home..........................105

6.2: Determining Your Requirements...........................108

6.3: Obtaining Accurate Bids 111

6.4: Getting the Plans and Permissions..........................114

6.5: Finding the Right House Plan 117

Chapter 7: Heating and Cooling   121

7.1: Types of Heating 121

7.2: Types of Heating Systems 124

7.3: Types of Cooling Systems 127

7.4: Air Cleaners and Humidifiers 132

7.5: Ceiling Fans 134

7.6: Fireplaces..............................136

Chapter 8: The Beginning of Your Castle (Phase One)   143

8.1: Permits, Building Codes, and Inspections.................................143

8.2: Groundwork and Excavation...145

8.3: Foundations, Footers, and Slabs 146

8.4: Underground Plumbing...150

8.5: Backfilling, Drainage, and Landscaping..................................152

Chapter 9: The First Nail (Phase Two)   157

9.1: First Things First..157

9.2: Frame Your House..158

9.3: Cap it Off: Roofing...162

9.4: Windows and Exterior Doors 167

9.5: Decks and Porches...171

9.6: Siding and Trim 173

Chapter 10: It Looks Like a House! (Phase Three)   177

10.1: Electrical and Plumbing..177

10.2: Insulate 181

10.3: Drywall and Trim...182

10.4: Finish Off the Kitchen 188

10.5: Finish Off the Bathroom 190

Chapter 11: Working the Walls   195

11.1: Flooring 195

11.2: Kitchen Cabinets and Counterpoise...............198

11.3: Bathroom Vanities..201

11.4: Finish Plumbing and Electrical.......................204

11.5: Stairways 206

Chapter 12: Lighting, Hardware, and the Accessories   211

12.1: Types of Light Sources.....................................211

12.2: Home Lighting Uses 213

12.3: Lighting Terms...217

12.4: Hardware: Doors and Drawers 220

12.5: Interior Priming and Painting.........................223

12.6: Appliances..225

Chapter 13: The Final Inspection   231

Chapter 14: Landscaping, Your Drive, and Walkways   235

14.1: To Pave or Not to Pave 235

14.2: What to Know About Driveways...........................236

14.3: Paving Materials 237

14.4: Walkways 239

14.5: Traditional or Xeriscaping 241

14.6: Ponds and Pools 242

Chapter 15: The Finish Line   245

15.1: Preparing for Taxes 245

15.2: Punch List..247

15.3: Moving in and Managing Your Investment............249

15.4: Setting up a Maintenance Plan 254

Chapter 16: Spending Now or Later   259

Case Studies 260

Conclusion   263

Appendix A: Interview Questions   267

Sample of a Subcontractor Suggested Interview Questions 267

Sample of a General Contractor Interview Questions 268

Appendix B: Checklists   271

Sample of a Financial Inspection Punch List................................271

Appendix C: Worksheets   277

Sample of a Budget Worksheet 277

Sample of a Lien Waiver 278

Sample of a Monthly Planning Worksheet....................................279

Sample of a Weekly Planning Worksheet 279

Bibliography   281

Biography   283

Index   285

INTRODUCTION

Congratulations! You are about to embark on the journey of your life. It will be exciting, frustrating, and tiresome. It can make you wish you never started, question your sanity, and drive you to tears, but building your own home will undoubtedly give you great joy and a sense of accomplishment, confirm your strength of character, and leave no doubt in your mind of your ability to complete any task you choose to take on.

This tome will assist you in determining the extent of your involvement — such as how much of the physical labor, or "sweat equity," you can contribute — and help to avoid the pitfalls, acquaint readers with the nuances of home building, and alert the potential home builder to the most efficient practices.

Think of yourself as an artist with a blank canvas, taking the first swipe with a brush and making the picture in your mind a reality. This book contains all the knowledge you need to make informed decisions, including advice from those who took on the same challenge. Building a home is not the easiest project you have ever tackled, but it will be the most rewarding — both emotionally and financially.

Constructing your own abode reflects your personality beyond the color of the walls or window treatments. If it has always bothered you to have the washing machine and dryer disturb your quiet office, it is your prerogative to arrange the room placement differently. Are your children's bedrooms too far from the playroom? How about the washing machine? Why is it always in the basement if all the clothes to be washed are upstairs? Tired of going up and down to do the wash? Change it! Your only limitation is your budget.

Think of the building process as a journey. Within the pages here are your detailed road maps to get you through the trip unscathed. It takes you home. Home is where we create our sanctuaries, our personal escapes, and the reflections of our individuality. Home is typically where we spend the majority of our time, where we can relax and be who we truly are. Because all these ingredients are so important in the integration of what a home is to each of us, it is equally as important for our home to be warm, comfortable, and inviting.

Knowing the process is important, whether you put every nail in its place, hire a contractor, or subcontract various aspects. It even helps if you are staying where you are but want to make changes. For example, let us say you want to put in a circular driveway. Deciding on a traditional cement or asphalt will take some preparation before speaking to a paving specialist. Forewarned is forearmed. Knowing what material to use and why will help you determine which contractor is best for you, as well as whether their selected material is one that you prefer, too.

> ### Notes From the Field
>
> It is inevitable to feel frustration as you experience the learning curve. Your inner critic will be working in overdrive, constantly shouting in your ear that you cannot do this. The reality is that you can. This book will help get you on your way to achieving this goal.

The main challenge of building a home is reducing risks before they occur. Knowing as much as you can when you start helps you avoid headaches and costly oversights while construction is in progress. There should be no surprises. It is much easier to prevent a fire than to put one out. This guide is intended to reinforce your confidence in the decision-making process and encourage you to discover that you are more capable of planning and putting plans into action than you initially may have imagined.

Included at the end of each section of this book are money-saving tips for at-a-glance reference. In addition to these tips, more money-saving ideas and tips are sprinkled throughout almost each chapter, further aiding you in your money-saving goals.

Many people walk out of custom home-building projects experiencing success and feeling a sense of gratification; others end up spending too much time, stress, and money on a project. Building your own home is a rewarding and risky adventure, and each of us walks into these situations with our own vision or an ideal home in mind. How it will all piece together is likely one of the most intricate puzzles you will ever construct.

Building your own home is one large task made up of numerous smaller chores, which initially seem overwhelming. We will demonstrate how to map everything out so you can understand the entire process without being overcome by a feeling of incompetency. In addition, by planning appropriately, you can keep yourself from blowing your budget. You will learn how to finance from the very beginning.

Notes From the Field

It does not matter if you have absolutely no experience when it comes to residential construction. It is possible to learn what is necessary to become a successful home owner-builder. Countless home owner-builders have proved this fact in the past, so do not let people try to convince you otherwise. Normally, those who have negative things to say are those who do not understand the importance of goal achievement.

CHAPTER 1
Getting Started

1.1: Why Build Yourself?

It is a good question — but only you can answer it. You are not going to be happy if you are pressured into the process. If convincing is needed, you are not ready. It takes an individual with conviction to do the job — a person who will not falter once construction has begun. The three most common reasons people build their own homes are:

1. To design exactly what they want and how they want it.

2. To save money by doing as much of the work as they can or are qualified to do on their own.

3. To have the experience and gratification of building themselves.

An owner-builder is defined as an individual or set of individuals who own the property where the home is going to be built and act as the general contractor for the project. This term also describes those who do all the work themselves and hire subcontractors to help build the home. To be considered an owner-builder, the owner-builder's work site must be his or her principle residence for at least 12 months subsequent to completion of the build — and in some states, even longer.

When people talk about building a custom home, others might assume they will be building the home with their own hands, but this is rarely the case. There are a number of contractors and highly experienced do-it-yourself-ers who have been known to achieve this goal from the ground up, so it cannot be entirely discounted. While the owner-builders can achieve a good amount of the work, the term is mostly derived from being their own general contractor. As discussed later in this book, it will be necessary to hire other contractors — plumbers, roofers, and electricians — to work on the house.

Here are some important questions to ask and seriously consider prior to making the decision to become an owner-builder:

1. Do I have enough job security to take off the time needed to build this home?

2. If not, can I work part-time or full-time in conjunction with building this home without burning myself out?

3. Is my schedule flexible enough to build this home?

4. What other obligations are present in my life that might pull me away from or delay this project?

5. Will I make money building my own home or lose money from the time not being at my job?

6. Is the construction process something I understand, or is it something I have a lot to learn about?

7. Do I have enough time and money to take classes about topics and other training necessary to build this home?

8. Am I good at managing people, time, money, and projects?

9. Do I understand and have an eye for quality?

10. What resources are available to me, and am I good at effectively finding them?

11. How are my problem-solving skills?

12. Does change affect me in a positive or negative way?

13. Am I an organized person, or do I rely on others for that?

14. Am I good with bookkeeping and accounting, or is this something I need to take a class to learn more about?

15. What do I stand to lose if this project folds, and am I prepared to be honest with myself about this?

16. Will I lose money, family, friends, or anything else if this project does not come to fruition?

Be honest with yourself. This will help you determine whether you have what it takes to be a general contractor or whether you need to employ one. These questions help you understand what you need to learn or brush up on to become one yourself. You might have plenty of confidence, but if you lack the fortitude to stick with it, the costs associated with being wrong could be significant.

The state where one builds will dictate the conditions under which an owner-builder may sell a home. Most states will require you to be a licensed contractor if you are "flipping" houses, a practice in which a house is bought and sold in a short period of time.

For those of you looking at the project with anticipation and a burning desire to start, you will probably have lots of fun. If there is any reluctance, you can be assured nothing will go right. Owner-builders need to get used to the fact that vendors may ship wrong items, wrong quantities, wrong colors, wrong styles, etc. Whether one builds for the fun of it, for the challenge, or out of necessity, the majority are proud of and pleased with their accomplishment.

If you ask these same individuals if they would do it again, you would be hard-pressed to find one who would not say "yes." These same individuals would describe how this was the best — and most gratifying — experience of their lives. The things that go wrong, primarily due to lack of experience and proper planning, are par for the course and rarely a reason for them to turn their backs on future projects. These individuals will most often chalk it up to being a good learning experience, which they can pass on to other future owner-builders.

Notes From the Field

It is not uncommon to hear an owner-builder discuss some regrets they may have, or how they would have done things differently if given the chance to do it over again. More often than not, this is because insufficient time was spent on the planning process. New builders should use these regrets as a lesson to learn from, instead of repeating mistakes and learning the hard way.

Money-Saving Tip:

Rather than attending many costly workshops and purchasing stacks of how-to videos, speak to other owner-builders who have already worked in the proverbial trenches. This will not only save you money, but it will give you an accurate first-hand account of what to expect when going through the building process. They can tell you the nuances of your local codes, vendors, and subcontractors. If you need confidence from videos, many libraries and video/DVD rental stores carry them; they can often be found in a few of the do-it-yourself departments of lumber yards and bix box stores for free or minimal rental costs. Your money should go toward building expenses, not buying an education you can get for free or at low cost.

For those who will not be satisfied without a formal education in the building trades, call your county education board or local school board to find out about adult education classes. Let your taxes work for you.

1.2: The Benefits of Building Yourself

There are numerous benefits to being an owner-builder. One of the biggest is getting a home that meets all of your family's needs without question. Aside from budget constraints, there will be no settling for anything. Instead, you will be choosing the floor layout and room sizes. Additionally, you will have full control of the decisions pertaining to everything from what kind of windows will be installed to the variety of hinges the cabinets will have; you will make virtually every decision about the home. However, this control can be intimidating to some people.

Aside from having complete control of your build, you will also save thousands of dollars. This fact is sometimes debated, but mostly by contractors and subcontractors who wish to remain in control of the work. Contractors charge 20 to 30 percent, depending on your reference source, of the home's proposed value as their fee. So, when people choose to build the home themselves or act as the general contractor, that 20 to 30 percent becomes instant equity. That is a huge motivational incentive to becoming an owner-builder. This will be discussed in-depth in Chapter 6: Financing Your Project.

There are two ways owner-builders can save money: through traditional sweat equity, and through cold sweat equity. **Traditional sweat equity** is working on as much of the build as you can. **Cold sweat equity** refers to overseeing contractors during the build, buying materials, and controlling the entire project. There are ways in which the roles can be combined and this, of course, will be discussed throughout the book.

Notes From the Field

Those who enjoy being in full control at all times will do particularly well in their role as general contractor. There is a caveat here. If you are not experienced in leadership, you will have to learn very quickly how to balance being assertive without being overly aggressive. Having a dispute with a tradesman in the middle of putting up sheetrock may well delay your project for weeks. Some skilled workers are not easy to find. The best advice is to know what you are demanding, including the implications. By all means, be assertive in what you want and do not let anyone talk you into something you did not plan, unless they make a compelling argument. In that case, check with other contractors for pricing and their opinions.

In addition to saving money, those who build their own homes have an amazing sense of gratification, prestige, and accomplishment. Building your own home is a huge undertaking, but the finished product is well worth the time and effort. Imagine what it will feel like crossing the finish line and finally moving in. Many would consider this the least of the reasons to become an owner-builder, but it is certainly the most rewarding, especially when you review the paperwork and photography collected during the course of the build. You will have a first-hand account of exactly what you and everyone else involved have experienced.

Document everything — not only for your memories, but also for the contractor's work. Use a camera to prevent insulting your workers when you think something is wrong. Contractors often hire workers, and you may need to demonstrate evidence if some individuals produce shoddy work, take frequent breaks, do not pay attention to what they are doing, talk on their cell phone while working, or even drink alcoholic beverages on the job. It will help to occasionally use a video camera, but it is not imperative.

On top of these benefits, you can select all of the building materials yourself and take the opportunity to add options such as shutters, brick walks, or even a basketball court. Do your homework to decide what you do and do not want in your home. Owner-builders have the opportunity to decide on everything from the type of heating system to the style, material, and color of the shingles you want. Rather than working with a standard set of plans and building materials, often regulated by commercial builders, this control is in your hands.

Being in command is a motivating factor for a lot of owner-builders, particularly those who cannot find a contractor willing to work outside their standard parameters. Owner-builders are only limited by their budget, so they are able to let their imaginations soar. The final result will be what you want, the way you want it. There is no need to argue over last-minute changes, as the owner-builder sets the schedule and there is a lot more flexibility. This kind of control is stressful and overwhelming to some, so be aware of that reality.

Being an owner-builder is an experience that will ensure quality, easy maintenance, and easy repair. It will also bring you closer to family, allow you to establish new friendships, and set aside funds for added comfort. There is the ability to create something as energy efficient as you desire, and establish a sense of personal growth that might not otherwise be achieved. You have the option to explore green building techniques that other general contractors may not be familiar with. We will discuss green building, but know that most general contractors build either traditional or green, and few hybrid systems. There is no limit to how much you can educate yourself.

People who build homes do so to suit their needs. They have usually gone through months — and sometimes years — exploring pre-existing homes and custom builders' plans in communities to no avail. Another consideration is the fluctuation of the home-sales market. Less expensive homes that have fallen into foreclosure have often equally fallen behind in maintenance. Bigger homes are not necessarily better homes, either. Yes, there are often great buys on large homes, but there are associated heating and cooling costs, as well as increased taxes based on the assessed value of the residence.

Remember, being an owner-builder is not for everyone; there is nothing wrong with facing this reality. The only wrong thing is putting yourself into this role if you know it is not what you are best-suited for. It cannot be over-emphasized: You must be honest with yourself.

Money-Saving Tip:

Speak to as many friends and family members as you can about your desire to be an owner-builder. Some will scoff, but do not be disappointed or deterred. Make a detailed list of who is willing to help you throughout the process, along with their skills. This is one of the most important steps in saving contracting dollars.

Those who have no time, skills, or desire to assist may change their minds when others participate. Show your appreciation for those who volunteer, and do not take a rejection as a personal affront. You do not know everything that is going on in their lives or their state of health. There is another consideration, though: Sometimes, plans change and volunteers who are not reliable can throw off your schedule. Have contingency plans should this occur.

1.3: Misconceptions About Building Yourself

Horror stories abound regarding what owner-builders go through when completing (or not completing) their building projects. But if you plan appropriately, this should not happen to you. When situations arise, keep a level head and stay confident. There will be highs and lows during the process, and you must maintain a steady nature. Rumors and misconceptions are rampant wherever people congregate to compare notes — the trick is figuring out what is true and what is false. Let us pick apart one misconception at a time regarding building your home:

Building your own home is more expensive than a commercial builder.

There are no concrete facts supporting this claim, no matter how many horror stories you hear. People might tell you it is common for an individual to have to stop the construction process because they went over budget, unlike if they had used a contractor. But this is rarely true because there is built-in equity from acting as your own general contractor. Proper planning will wipe out this misconception before it ever becomes a reality.

Owner-builders are less likely to pass an inspection than contractors.

Owner-builders can be more meticulous when it comes to what the inspection entails because

1. They will be living in the home

2. The obsessive-compulsiveness of getting things done right the first time is present

3. Inspectors know from first-hand experience that a contractor is more likely to try to dodge a building code or cut a corner than an owner-builder. Owner-builders do not know what they can and cannot get away with, so they are more likely to follow every rule and detail to the letter the first time around.

You will be paying more for supplies, rather than getting discounts through a contractor.

Many will tell you this should top the list as one of the biggest misconceptions of being an owner-builder, but with effort, research, and persistence, it is possible to get prices as good as or better than the prices contractors pay. You just have to know where the deals are, whom to talk to, and what to say. Learn not to settle until you are satisfied. Shopping around from supplier to supplier will yield the best results. You will also find suppliers who will match or do better than the written bid you show them from another supplier in order to obtain your business.

You cannot build a house as nice as a contractor.

Owner-builders may have the ability to build a home that is nicer than a contractor's work because it is going to be their home, and it is going to be exactly what they want; it will not be based on another person's idea of what a nice home is. Owner-builders have the ability to look at floor plans they like, not what someone else's definition of what they should like. This reality wipes this misconception out before it even becomes a conversation.

You must have a license to build your own home.

If you built a home that you did not intend to live in, this claim would be true. It would also be true for those who build and sell homes for a living. Use your state or local government Web sites to research what you need a license for, such as plumbing and electrical. It entirely depends on the building codes. If you are new at electrical work, then it is in your best interest to obtain the services of a licensed electrician to hook up your circuit boxes.

It is not legal to be an owner-builder.

There are no laws anywhere in the United States stating you cannot build your own home. As long as you follow all the same guidelines required of a licensed contractor, there is nothing to worry about. The only illegal part of being an owner-builder is if you fail to follow building codes — the house would then fail inspection, so it is not a wise move. Besides, who wants to live in an unsafe building? That is why building codes exist; they are not put in place to employ certain trades.

Notes From the Field

Practically speaking, be prepared for negative comments from those you might have expected to support you. Some people are traditionalists and have little or no spirit of adventure. This is your project, not theirs. Achieving your goal will take more than financial and physical needs; it can be emotionally challenging if you let the negativity of others affect your achievement. It takes a great deal of foresight to meet your goals, and it does not matter whether others think you are smart enough. Let the finished product speak for itself.

Misconceptions and bad advice can ruin relationships as well as dreams. As much as you trust and honor the opinions of those you ask, always do your homework. You can ask two people what kind of window design they like and get three different answers. The same applies to building techniques. The people you should speak with are professionals, as well as those who have gone through the same process. Some of the lumber companies and big-box stores hold seminars for owner-builders where you can meet people who are trying to make the same decisions. If someone offers discouragement instead of support for your decision, analyze his or her response objectively rather than discount or follow it.

Who you ask is just as important as what you ask. It is fine to ask your 16-year-old brother what color to paint the front door, but heed the advice of experts when it comes to what kind of paint you use, if at all. For example, if you are living in the high desert of New Mexico where the winds frequently whip up sand storms in warm weather, exposed painted doors may last a year or two given the sandblasting effect. You may have to use a stainless steel entry or build a protective wall. Only someone familiar with the area or these conditions would know enough to clue you in.

Another consideration is where you get your information. The Internet has become a favorite source because it is generally free; however, sometimes you get what you pay for. A vendor of a product will do what they can to sell their wares. And while most manufacturers are honest, some may be misleading. If a site does not tell you the negatives uses for a particular product they manufacture, look elsewhere. Sticking with the paint example, there are various kinds of paint, and the most popular are latex and oil-based. Some are washable, some are not. There are advantages and disadvantages to each. Does the Web site explain the difference? Do they tell you to use one coat or more? Is it a bargain if you need two coats versus a more expensive one coat?

Some good Web sites are those of handyman shows aired on radio and television. Glen Haege happens to be a favorite of many, but you will have to check with your local radio station for time and day of his broadcasts. He also has a Web site at **www.masterhandyman.com**. Such handymen answer questions, compare products, and have reputations to protect — so while they may have sponsors, they are quick to point out their relationship.

Published books, pamphlets from professional organizations, and information from a local technical school are all good sources of advice you can rely on. In fact, some cooperative education schools offer short classes for those entering the construction field.

Scheduling is another concern. If someone has not been through the design and construction of a new home, he or she might have no idea how long it really takes. A suggestion regarding a schedule to someone building their own home would be to ask them to create a schedule, and then tell them to double the amount of time that they have put aside. Most inexperienced builders do not understand all the issues that will come up — not only during the design phase of the project, but also during construction itself.

One of the biggest reasons people give for why they want to build their own home is to save money by doing it themselves. But do not cut corners. Downloading house plans, and doing your own painting, hammering, and clean-up are cost-saving. However, anyone can draft certain legal agreements by accessing the Internet and downloading the appropriate forms. If they do, this does not mean that they have put together a good legal agreement. Without the experience of a good attorney, they may not understand the mistakes they have made with their agreement until it is too late, when the contract has been violated and an attorney will be needed. We discuss this in more depth in our last chapter.

There are numerous issues that arise during construction that no "how-to" book will prepare you for. One of these issues has to do with hiring subcontractors to finish plumbing, electrical, or HVAC (heating, ventilation, and air conditioning) work. The good subcontractors work with eight or ten other general contractors. Home owners are their "bread and butter," and each needs to understand what the other needs so that work goes efficiently. An inexperienced owner-builder does not have this same relationship. More than that, he or she may need assistance getting a project ready for the appropriate subcontractor. This means there is a learning curve for owner-builder and, if subcontractors are busy, they do not want to work with inexperienced people because it can take them away from their bread-and-butter customers. If this is the case, getting key subcontractors not only to the work site but also to just return calls can be a challenge. This is just one source of frustration that a novice builder will encounter.

Due to issues like this, any schedule is difficult — if not impossible — to follow. Too many things are out of the new builder's control, and he or she needs to understand myriad delays that can take place. But the sole motivation behind being an owner-builder should not be cost-savings. Do it instead for the experience. Do not rush, and take the time necessary to evaluate each step of the process. It can be a charming experience as long as budget and timing are realistic.

Money-Saving Tip:

The Internet is your friend. Do not forget this. Write it down if you have to. You will hear that a lot throughout these pages. Use the Internet to connect with other owner-builders on forums and chat-type Web sites. This will save you money in that they will be able to point in the direction of what is true and what is not is because they too are going through the building process.

CASE STUDY: DAVID LUPBERGER

ServiceMagic Home Improvement Expert

Golden, CO 80401

www.servicemagic.com

"Building your own home has a romantic quality to it. The thought of designing, building and living in your own "shelter" is an appealing one. There are so many things to consider, and that can be a source of both creativity and frustration. In building your own shelter, there are hundreds, if not thousands, of design decisions that need to be made. Most of these decisions affect the budget, so obtaining an in-depth checklist of design decisions is an important starting point. A good plan can take a year or more to evaluate, and this process should not be rushed."

CASE STUDY: DAVID LUPBERGER

Scheduling is another concern. If someone has not been through the design and then construction of a new home, they will have no idea about how long this really takes. If I was to make a suggestion regarding a schedule to someone building their own home, I would ask them to create a schedule, and then tell them to double the amount of time they have put aside.

Most inexperienced builders just do not understand all the issues that will come up not only during the design phase of the project, but also during construction itself.

One of the biggest reasons I hear people mention when wanting to build their own home is that they can save money if they do it themselves. I always tackle this statement head on. For example, anyone can draft certain legal agreements by going online and downloading the appropriate forms. Doing this does not mean that they have put together a good legal agreement. Without the experience of a good attorney, they may not understand the mistakes they have made with their agreement until it is too late.

I bring up this example because there are numerous issues that arise during construction that no 'how to' book will prepare you for. On of these issues has to do with hiring subcontractors to finish plumbing, electrical, or HVAC work. The good subcontractors I used to work with also worked with 8 or 10 other general contractors. We were their 'bread and butter', and we each understood what the other needed so that work went efficiently. An inexperienced homeowner/builder does not have this same relationship. More than that, they may need assistance getting their project ready for the appropriate subcontractor. This means there is a learning curve for homeowner/builders, and if subcontractors are busy, they do not want to work with inexperienced people because it can take them away from their bread and butter customers. If this is the case, getting key subcontractors not only to the worksite, but also to just return calls can be a challenge. This is just one source of frustration that a novice builder will encounter.

Due to issues like this, any schedule is difficult if not impossible to follow. Too many things are out of the new builder's control, and they need to understand the myriad of delays that can take place. I always try to tell aspiring novice builders to not do their own house to save money. Do it instead for the experience. Do not rush, and take the time necessary to evaluate each step of the process. It can be a charming experience as long as budget and timing are realistic."

1.4: Your Role as the General Contractor

Being an owner-builder means wearing many different hats, including that of a **general contractor**[1]. The skills necessary to assume that leadership mantle mirror that of an entrepreneur or a company president. You must be personable because there will be internal conflicts to resolve. Juggling numerous projects while maintaining a tight schedule is your job. It will be necessary to understand the construction process in order to ensure everything runs safely and smoothly. This role mandates being good with money because you are also the accountant, buyer, and budget keeper.

The ability to see the big picture is not an option but, rather, a requirement. If you can look at the house plans and visualize what needs to happen next, then this is the right position for you. Conversely, if you look at the house plans and see lots of lines in a pretty juxtaposition, you might benefit from enlisting outside help. Be as honest with yourself as possible. Your vision, or lack of, is going to play a large role in your owner-builder project. If you are not honest about your abilities, this entire project could end up a big mess, costing you thousands of dollars in materials and manpower.

This is no small task, to say the very least. There is much involved when deciding to become your own general contractor. Having a firm understanding of the entire process early on, though, will alleviate stress and potential mistakes made along the way. Reading this book cover-to-cover is a good first step. Here is a list of some of the various roles you will be playing as the general contractor:

1. **You have hire-and-fire power.** That means on any given day you are at the job site, you have to be prepared to get rid of anyone who is not getting the job done and replace them with others who will do it

---

[1] General Contractor is the team leader, cooridinating the actions of all others

right. This leadership ability is extremely important and will prevent things from going wrong on the work site. You have to be constantly aware of everyone, what they are doing, when they are showing up, and when they are leaving. To some, this is a hard role to play because it may involve uncomfortable confrontation. Remember, this is your house, and only the people working to your standards should be working with you.

2. **You will be a scholar.** To be a general contractor, you have to have some knowledge of residential building. Learning the ropes ahead of time and staying brushed-up on various topics throughout the build will keep you on top of things and will help avoid costly mistakes. There are numerous classes and workshops available about home building, home financing, and home ownership. You can obtain lists of classes from your financial institution, a well as on the Internet.

3. **You will be the communicator.** Talking on the phone, giving orders, and having meetings will be daily duties as a general contractor, so the ability to communicate well is a crucial skill. Be clear during each and every exchange about what you expect, whom you expect it from, and when you expect it. This will alleviate all confusion and miscommunications. Be as firm as you are clear. Having confidence in yourself and also your project will help with this. If you are the shy type, it will be necessary to break out of your shell; otherwise, things will not be accomplished and you will not be taken seriously.

4. **You will be obsessive.** Every little detail will matter to you. If every detail is accounted for every time, your house will be exactly what you want it to be. This is where an eye for quality will come into

play. If you notice something that is not the quality you are looking for, it can be changed. Do not let other contractors or subcontractors argue these points with you. Remember, you are paying them to do the job you want. If you are the type who tends to overlook details, employ the help of a nit-picking friend or family member.

5. **You have to be more dependable than anyone else.** It is important for you to be on-site on a daily basis for a good chunk of the day. You should be calling everyone else asking where they are rather than vice versa. If you say you are going to be there at any given time, be there. If you cannot be there on a daily basis, employ others to do so. This could be friends, family members, or co-workers who pop in regularly to oversee what is going on and report back to you.

6. **You will be the problem-solver.** As general contractor, you have to solve every problem — including reordering the shower doors that arrived broken and the molding that came in the wrong color. Passing the buck is not an option here. If the problems become too numerous, delegate tasks to a spouse, a family member, or a trusted friend. It will still be your responsibility to follow-up on these issues and ensure the delegations are happening the way they should.

7. **You will be an accountant.** There will be plenty of money exchanging during the entire process and, as general contractor, it will be your responsibility to make sure everything is being spent efficiently. You will also be responsible for staying on top of the budget, revising the budget when money starts running low, and making necessary changes as needed. Attention to this matter is a top priority; otherwise, the entire project will fold. If you are not good at bookkeeping, find someone who is.

8. **You will be a shopper.** It will be your responsibility to make sure all the elements in your home are absolutely right. Shopping for the right items, right prices, and quality are priorities. If you are not careful, you will end up with something that is not exactly what you want. This situation is not favorable and can be avoided. It is easy to go overboard when you are buying things for your own house, but let your budget guide you — not your impulses.

You will be amazed at all the little things that will continue to pop-up throughout the building process. Everything from deciding colors, understanding flooring options, talking to inspectors, hauling off trash to the dump, or calling a clean-up crew will be part of your job description. This does not even include time management. You are required to perform many roles; wear these hats well, and you will be successful.

Still, remember to try to have fun whenever possible throughout the course of this owner-builder adventure. Fun is normally the last thing discussed when the topic of hard work is on the table. This is your building project, though, and there is no reason why you cannot make it as positive and as enjoyable an experience as possible. This is not only beneficial for you, but for everyone else who is going to be part of this building project.

Above all, experts all agree that your schedule must be flexible to ensure a better success rate. If you possess the type of temperament that flies off the handle each time an appointment falls through or someone shows up unexpectedly, it might be best to enlist the help of others to help manage the workers while you keep a level head throughout the building process. There are no ideal owner-builders, but if you work seasonally, are retired, or have enough free time to manage the people and process, then you may be ideal to be your own general contractor. It helps to have certain skills, but if you are willing to learn, you can obtain the knowledge.

Money-Saving Tip:

Purchase some financial software to help you stay organized throughout the building process. There are a number of accounting and financial softwares to choose from, and Web sites like *www.cnet.com* will compare the products for you. If your finances are not organized, you will lose money. If using a computer does not suit your lifestyle, purchase budget-keeping notebooks. These are found in office-supply stores and many superstores.

1.5: Finding and Managing Subcontractors

Whether acting as your own general contractor, hiring one, or doing the whole job yourself, it is likely there will be times you are going to need the assistance of general subcontractors. Be prepared for a prolonged selection process. It can be difficult finding good subcontractors who are affordable, dependable, and reliable, all while providing quality workmanship. This is where the term "caveat emptor," or "buyer beware," comes into play. There are numerous risks involved with working with bad contractors. Work with subcontractors you know through referrals and recommendations whenever possible. Lenders and building-material suppliers are a good source for quality references.

Be as selective as possible when assembling your team of contractors. These will be the people erecting the home you and your family will be living in for a long time. Selecting the most inexpensive may also turn out to be the cheapest in quality, and that will not be in your best interest, so keep it in mind at all times. Like with everything else, you generally get what you pay for. Here is a list of types of subcontractor you might need to complete your build:

- Excavator
- Foundation contractor
- Framer
- Brick, siding, or stucco contractor

- Electrician
- Plumber
- Roofer
- Insulator*
- Drywall hanger and finisher
- Painter *
- Flooring contractor
- Cabinet installer

- Ceramic tile installer
- Asphalt contractor
- Trash hauler *
- Well-driller if town hook-ups are not available
- Septic tank-installer if town hookups are not available
- Landscaper *

Money-Saving Tip:

Sometimes the excavator works with his or her own foundation contractor for a reduced rate, so check out this money saving option.

Finding each type of subcontractor is as easy as attending home and garden shows or joining a local homeowner's association. This is a good example of trying to find the fun and enjoyment during the owner-builder process. Turn these events into an adventure, a scavenger hunt, or a treasure hunt. Use your imagination and make it a memorable experience, not just another errand to run in your list of things to do.

Having discussions with building suppliers or lumberyard workers is also beneficial to finding reputable subcontractors. The interview process can be daunting, but it is necessary. Write out a list of questions and other relevant information in a computer word processing program. This allows for easy inputting of information, and ideas can be moved easily from one file to the next. Act like you are a private detective trying to solve a crime or a reporter trying to land the next big scoop to make things fun and interesting.

* **You may be able to do these tasks yourself.**

Try conducting each interview over the speaker phone so you can type as the interviewee answers questions. Another option is to e-mail interview questions and request that they respond using e-mail as well. If you do not mind doing transcription, you have the option of tape recording interviews to type out later. You will find sample interview questions at the end of this book in Appendix A: Interview Questions.

Notes From the Field

Remember that these contractors are just as busy as you are. Do not have long-winded conversations with them. Keep things as productive as possible by keeping your questions and comments short and to the point while remaining cordial and personable. If you need practice, do so with a friend, with a family member, or in front of a mirror — it may sound silly, but it works.

A note to the wise: Once the contract is signed, they are your employees. These are professionals, and they do not need to be reminded that you are in charge. Develop a good working relationship as you would with co-workers on your daily job right from the start. Here are some stress relievers:

- **Be prepared.** There will be competition for your work, so be prepared during this process with schedules and contracts. Each subcontractor should present a bid on each proposed schedule (short, medium, long); that way, you have all the information needed to base your decision on, and you will avoid unexpected charges.

- **Expect changes.** After the schedules have been discussed and contracts have been signed, there will inevitably be changes throughout the build. You might not like the windows you originally chose, or you might change your mind about what kind of sinks are in the bathrooms, or you might want ceiling fans in each room. These changes must be reflected in the schedule to prevent you from going over budget.

Good leadership skills are essential to successfully managing subcontractors. Keep the environment as positive and friendly as possible, and offer praise and encouragement. The last thing anyone wants to hear day after day is that they are not doing things right or their opinions are wrong. Be objective, watch how you say things, and try not to come off as a "know-it-all." Not only will the quality of the work be what you are expecting, but it will also keep the subcontractors coming back every day until their job is finished.

Money-Saving Tip:
Rather than holding meetings in restaurants, which can get costly, set up some coffee and snacks at the building site. Hold all of your meetings there so the contractor can see where they are potentially working and you will not have to foot the bill at the restaurant dozens of times. It not only saves money, but also time.

1.6 Going Green and Environmental Considerations

Rather than being a buzz word or political topic, many people see building environmentally sound residences as a worthy investment. It is especially applicable if you or anyone in your family has allergies. This applies primarily to the interior, such as carpeting. But insulation is sometimes blamed for allergens as well. You can either partially or fully build green, starting with the kind of construction you are doing. Depending on where you live, you may well be familiar with forged earth homes, adobes, and log homes.

Environmentally sound building involves factors such as:

- Low-flow plumbing fixtures
- Collection and storage of rain water

- Drought-tolerant landscaping
- Land use
- Energy efficiency of appliances, utilities, and renewable energy, which is fuel that can be replaced, as in solar power, ethanol, and passive heating by design
- Insulation of the building envelope
- Selecting materials from manufacturers who plan end-of-life-cycle disposition (renewal)
- Durability and performance of building materials
- Selecting indoor materials that have low emissions of chemicals
- Adequate ventilation

The idea is to use renewable resources, employ the most energy efficient means, and use recycled materials as often as possible. There are many products now made from recycled items. One of the most impressive is recycled tires mixed with cement in T-blocks[3]. They are utilized in:

- Soil erosion control and slope protection
- Wetland reconstruction
- Flood control
- Dike and levee construction
- Sub-foundation stabilizers, in particular where vibration is a factor

The latter is probably a good consideration if you live in a quake-prone region, such as California. If you are going to be building in a rural area, the question arises whether you will be on or off the electric grid. Will you be self-sufficient with solar electric? It really does not matter because you can plan

---

[3] T-blocks are an innovative blocks used in construction. Some are traditional squares with or without holes and provide greater stability than conventional commercial products. Other t-blocks are used to connect corners of walls and have found a place in constructing quake-resistant structures. The product is evolving rapidly to many uses.

on solar electric and heating even if you live in the middle of Chicago. Will you depend on the local water supply or dig your own well? Waste is an environmental issue, as well. Most municipalities specify the requirements for toilets, although in rural areas you may have septic tanks or even cesspools. Environmentally, there are new and green-friendly septic systems that do not require the bio-hazardous trenching and worries of the past. The problem is that not every septic system installer will tell you about the alternatives if they are not familiar with them. As with everything else, be educated.

Some green building products may be initially more expensive, but in the long run will have an impressive return on investment through energy-efficiency savings, as well as increased resale value of your home. More of interest to you at this point, though, is the number of financial incentives given by utility companies and manufacturers, plus local and federal governments, when you go green.

For more information on environmental building and community involvement, see **www.BuildingGreen.com.**

Getting started means being prepared for problems and errors. It means planning for the unexpected and having the cash reserves if necessary. Litigation may come later to recover the expenditure, but you will need the money up-front to pay other workers.

CASE STUDY: PATRICK LISKA

Liska Construction Company

Verona, NJ, 07044

Patrick Liska was working on a project where a drainage pipe was to be run around the new house. This pipe connected all the gutters and leaders from the roof, as well as a foundation drain. The surveyors had the drains ending in a stream not far away. However, when Liska used his transit and shot the height, the distance was not the same as the ones the survey team gave him. How the initial error occurred is unknown, but the result was they required additional pipe to reach the appropriate point in the stream, some 30 feet further. The error was perhaps considered minor but the cost was significant for both labor and material.

Did the surveyor have insurance coverage for his mistake? Who paid for it? These are the kinds of problems you may not prevent; but solid contracts assuring prompt payment is essential to maintain your financial viability.

CHAPTER 2
Preparing for the Process

2.1: Planning and Organizing

If there is one thing we want you to take away from this chapter, it is this: Above all, the most vital part of the construction process is not the construction, but rather the planning and preparation. Once you come to terms with this, you will realize more positive results as an owner-builder. You will also experience less frustration.

Whenever taking on a multi-step project, there is always the risk of something going wrong. The need to be highly organized is imperative when it comes to being an owner-builder. If you are not an organized person, you will benefit more from the material presented in this chapter. But do not become intimidated by these challenges. Find someone in your family or a friend who is organized to help you through this stage.

The first thing you need to do is create a plan, which will save you time and money. It costs nothing to create an advanced plan, and it causes no delays. When it comes time to implement the plan, you can work at a more relaxed pace. It also relieves stress when you and your crew know exactly what is going to happen and when. In this plan, make the following determinations:

- If you are a licensed contractor and have built multiple homes, then there is no question to how qualified you are to act as a general contractor.

- If you are like the rest of us, though, determine how much of the construction work you are capable of doing. Factoring in what you are sure you can accomplish will have a positive impact on your plan. Avoiding what you cannot accomplish will help steer you away from a negative effect on your preparation.

Your next task is to create a dream-home notebook. This notebook will contain your personal thoughts and observations related to what you want for your home. This can later change to a photo album, a three-ring binder, or a word processing file. The point is to keep your thoughts organized, maintained, and updated so they can be later implemented into your building plan. Choose a format you feel most comfortable working with because it will be a tool used often throughout the building process.

Following that, you have to set up a system to prepare for all the paperwork you are going to accumulate throughout the process. Because each step of the development generates its own set of documents, you will want to be prepared in advance. You do not want to lose crucial documents you will need later. A typical construction project will generate enough paperwork to fill a portable file carrier, the kind that latches closed and can be carried with you. Here are some suggestions for file tabs:

- House plans and design
- Correspondence with subcontractors

- Bids from subcontractors
- Contracts
- Bank documents
- Invoices paid
- Receipts from subcontractors

- Purchasing of property
- Materials spec sheets
- Permits
- Warranties

Notes From the Field

If this filing system is not large enough to accommodate all the materials you are collecting throughout the building process, do not be afraid to expand. Some owner-builders have reported using five-drawer filing cabinets and filling them to the brim with paperwork. Though this may seem excessive, it is not uncommon for mass amounts of paperwork to pass back and forth between you and other individuals and businesses. The type of filing system must work for you; otherwise it will not be utilized properly.

Once the filing system is in place, you will want to create a portable system that can easily be changed out and updated. Purchase a three-ring binder and fill it with dividers. As paperwork comes in, decide if it needs to go in the notebook to bring with you to the building site or if it needs to be filed into the more permanent system described above. If it should be brought along to the site, photocopy it and place it into the appropriate section of the binder. Each day, either at the beginning or the end, review the contents of the binder and either add or omit any necessary documents.

Money-Saving Tip:

Time is money. Do not let yourself fall out of good organizational habits. Disorganization not only wastes time, it also wastes money. The last things you want to do when working on a tight schedule is hunt for paperwork or worse, lose paperwork. You should be able to grab and go each day. Be conscious of where you are setting things down, where you are placing things in your vehicle, and where you are writing phone numbers. If you start falling into bad habits, you may find yourself in needless trouble.

2.2: Budgeting Your Project

The first thing you have to understand about the budgeting process is that this budget will continue to change throughout the entire build, much like the schedules and timelines. It is vital to accept and adapt to these changes and, most of all, expect them to happen frequently. To get yourself started, create a preliminary budget. Balancing costs with affordability can be a tricky process, but it is not impossible. Picking apart the process will help put it all together into a workable budget.

The first things you need to look at are your finances. Do this with your lender so when questions and concerns arise, a professional will be there with the right answers and advice. Here is a list of items you should to take into account during the budgeting phase of your project:

- Do you have any cash on hand?
- Do you have any issues with capital gains?
- What is your current tax bracket?
- How long do you intend to live in the home?
- Should you develop a long-term investment plan?
- What is the property's appreciation or possible depreciation?
- Are there tax deductions for interest and points?

By answering these questions, you and your loan officer should be able to calculate a loan with a payment you are comfortable with. If additional documentation is needed, your loan officer will ask you to bring it with you to the initial or second meeting. You may also have the option of faxing information to their office and following it up with a phone conference.

To develop this budget further, take out the bids you have collected from subcontractors. This information will provide you with a construction budget. This information is based on your specific plans, so nothing should be left out. Input all this information into a spreadsheet for quick reference. An example of a budget worksheet can be found in Appendix C: Worksheets.

As you can expect, changes in the actual costs in comparison to the bids can alter the totals considerably. These changes also modify the equity as well, so it is significant to keep the bids as close to or below budget whenever possible. This spreadsheet will act as a preliminary tool when setting up your budget, then will act as a tracking tool as costs are assessed and actual expenses incur throughout the build. If you are not interested in working with technology, graph paper will work just as well (write in pencil).

Notes From the Field

A mistake new owner-builders tend to make is not accounting for unexpected changes in their budget. You have to be realistic about the fact things can go wrong and these mistakes may cost money. If you plan everything down to the penny, you are being very organized, but very unrealistic. Pad everything, else you are shorting yourself too much. These issues will cause you delays, which will end up costing you money.

There are four keys to a good owner-builder budget plan:

1. It can be easily updated on a computer or changed on paper.
2. Each estimate is based on the bids provided, so they are accurate and specific.
3. Nothing has been omitted.
4. It contains a cushion for unexpected expenses that come up along the way.

Now that you have an understanding of what a spreadsheet should contain and what it might look like, create a worksheet either on your computer or on paper. Create columns that have a description, budget amount, actual cost, subcontractor/supplier paid, and contact information. This information can be edited to suit your own needs based on your specific build. When you are finished, the budget worksheet could be up to ten pages of written or typed information. That is how detail-oriented the planning process must be, every step of the way.

Estimating a budget early will help you plan for changes and make easy modifications. Remember that the Internet is your friend and will prove to be an invaluable resource tool. Go to your favorite search engine and type in the phrase "building cost estimator." You should find excellent resources and leads in your results.

Money-Saving Tip:

When securing bids from contractors, ask them for a copy of their "Take-off List." This is the materials list they will be working from and basing their bid on. Lumberyards will price materials out on these lists, or you can do the homework yourself. Visit at least three lumberyards, if possible, and create a spreadsheet to compare prices. You will learn quickly that prices differ considerably from one yard to another. Create material orders based on the lowest prices in the chart.

2.3: Hiring a Contractor

Can you do this?

If you decide not to be an owner-builder, do not worry. It is common to be anxious about this decision, and there are plenty of talented and professional general contractors available who can get the job done. Finding one who will build your home exactly the way you want might be a challenge, but it is not impossible. Many general contractors do not like to stray from the sets of house plans they work with or the set of subcontractors they work with. Conversely, it is not uncommon to find someone who will.

The role of the general contractor is to oversee all the work happening at the building site, manage all the subcontractors, and order all of the materials. In most cases, the general contractor requests a contract to be signed and payment be made only to him, and the contractor will be responsible for disbursing payments to subcontractors and suppliers. All communication about the job should be between you and the contractor, then they will relay this information to the subcontractors.

Notes From the Field

A general contractor will free up a lot of your time, as well as stress. If you are not the type who is good at managing many team members, then working with a general contractor will be for your benefit. Even though this individual will be managing all of the team players, he or she will still take instructions directly from you. If you hire a good general contractor, his or her first concern will always be making sure you are happy with the way things are going. This alleviates angst and relieves stress.

The Hiring Process

Finding a good general contractor is not easy and requires that you do your homework. Ask for referrals from friends, family, co-workers, and respected businesses in the area (the bank you plan on going to for financing, for example, might have a list of referrals available). In all cases, no matter how the contacts are made, follow your instincts. Well-meaning people you ask may know a contractor but be unaware of their work quality or ethics. Use referrals as a starting place, not the end decision.

As with choosing subcontractors, it is wise to find about three to five general contractors to secure bids from. This will allow for the same line-by-line comparisons and ensure that you are getting the right individual for the job. When conducting interviews, in-person if possible, present the same detailed information you would to the subcontractors. This will allow for a clear understanding of what you want, what kind of schedule you are looking at, and your budget for the entire project.

Because you are going to be establishing a working relationship with this individual, compatibility is crucial. You will be working with each other daily; therefore, it is vital both of you can communicate and get along well. Remember that first impressions can sometimes be deceptive because general contractors know they are increasing their marketability by being as friendly as possible during the first several interactions. During the interview, ask the contractor the following:

1. Can you provide a list of references? (Ask for between three and five)
2. Have you worked on any projects similar to mine?
3. Do you have a project portfolio you can show me?
4. Do you have a crew, or do you work from a stable of subcontractors?
5. Will you be willing to work with subcontractors I have chosen?
6. Will you be working on this project exclusively?
7. If not, how many job sites are you normally on in a given week?
8. How much time can you dedicate to working on-site daily?
9. Will everyone on-site carry the proper licenses and insurance?
10. How do you work with customers to save time and money?

Be sure to create a set of plans and drawings to provide each contractor either before or at the time of the interview. Review every detail of these plans with each contractor so they are clear about your expectations. If you feel they are not being honest about time constraints or budgeting, ask pointed questions that will help lead you toward the right decision. If they give you any indication at all that they do not need to see your plans and schedule during this meeting, move on to another contractor. This is a clear red flag.

While the contractor prepares the bid, contact all their references to conduct interviews. If possible, visit some of the sites they have worked on to inspect workmanship and ask questions about reliability. When it comes time for the bid to be delivered, request that it be broken down by specific tasks into a fixed bid so you know where your time and money are being invested. If it is a time and materials bid, costs have a way of escalating out of control quickly, and this is not a good situation for anyone to be in. Below is a list of ten common mistakes made when hiring a contractor:

1. Not having a clear understanding of what you want (bad planning)
2. Not getting everything in writing
3. Missing dates in the contract
4. Paying a large deposit upfront
5. Hiring unlicensed or uninsured contractors
6. Hiring the first contractor you speak with
7. Not understanding the delays that will occur, such as with weather and illnesses
8. Expecting things to be neat and perfect all the time
9. Not listing penalties in the contract to protect the homeowner
10. Thinking that to have a contract will prevent problems or dishonesty

The contract phase of the agreement is another significant exercise in good communication skills. Ask pointed and detailed questions about everything you do not understand and do not agree with. Otherwise, you might have a negative experience, or your expectations might not be met to the degree you desire. Some elements of a good contract include:

- A start and finish date
- A statement that lien waivers will be provided for the contractor, suppliers, and all subcontractors
- Payment schedule (no money, unless it is earnest money, should be paid up front): typically, 30 percent paid at each milestone completion, then the last 10 percent when the job is completely finished.

For more about lien waivers, see this Web site: **www.infoforbuilding. com/Lien_waiver_N.html**. An example of a lien waiver can be found at the back of this book in Appendix C: Worksheets.

Before signing anything, be sure you are completely satisfied with each point of the contract, that the contractor has a full understanding of the job, and you feel comfortable with the rapport you have with the contractor. If you are working with a professional, he or she will try to accommodate everything you are looking for in the build and be honest with you about what is not a realistic expectation. Be open to changes and delays, and trust that the work will be accomplished.

Money-Saving Tip:

Although it is unconventional, and most general contractors may not agree to this, ask if you can do all the material purchasing. Many general contractors inflate the cost of materials by 10 to 25 percent for various reasons. You can save that cost if you are making the purchases and arranging for deliveries (or picking up the materials yourself). It does not hurt to ask, and if you feel strongly about this aspect of the agreement, you can add it into the interview process.

2.4: Dealing with Insurance Issues

Even though you are the owner-builder, you will still need insurance in place to complete this build. When dealing with a lender, certain insurance may be required, including:

- **Liability:** This protects the homeowner against anyone getting hurt on the property during the building process.

- **Workers' Compensation:** If the contractors have employees, they will have this in place. If they do not have employees, the subcontractors will fall under the liability insurance.

- **Course of construction:** This insurance, also referred to as Builders Risk Insurance, protects the homeowner from theft, fire, weather, or other damage during the building process.

- **Payment Bond:** If your subcontractor should default or vanish during the construction process, your lender may require this bond be in place from your insurer to provide payments to suppliers.

- **Disability insurance:** This coverage will provide compensation if you get hurt on the site and are forced to miss work as a result.

- **Term Life:** Some lenders require that this extra coverage be in place on you and your spouse, if applicable, with the bank listed as the beneficiary on the policy should you die during the course of the construction process. This will give them necessary protection and allow them to recover their investment.

The financial institution will provide the type of insurance necessary during the building process and all of its requirements. If you ask, most often the lender will have a list of providers available for you to call. Shop around for the right terms and rates before making the final decision. This is another aspect of the process that requires a lot of lead-time. Give yourself plenty of time to make and receive phone calls, and schedule appointments.

Purchasing a standard homeowners' insurance policy is one way to offer protection during the build. How much coverage you need (or are required to have) can be easily determined while price shopping. This will cover any damages that would be covered by the liability insurance discussed previously. Another option is to purchase insurance that provides coverage to the physical structure rather than the liability. Some mortgage companies, though, require the standard policy to be in place in order for the financing to go through. Whatever policy you decide to go with, evaluate it once the build is complete to be sure you are receiving the proper coverage.

Notes From the Field

Owner-builders have reported big problems associated with failing to ensure proper insurance is in place prior to starting the building project. Issues have ranged from injury-related lawsuits to unresolved thefts. These nightmares are preventable if you take the proper course of action from the very beginning. You cannot afford to take the risks from skipping this process.

Insurance issues for owner-builders include the possibility of paying workers' compensation and also liability costs if one of the workers gets hurt on the job. The best way to ensure you, your general contractor, and your subcontractors are in compliance is to speak with your financial institution and the agent from whom you are purchasing your homeowners' insurance. You will also want to inquire about insurance covering materials and equipment stored on the building site.

Money-Saving Tip:

Be sure to talk to the insurance agent if there are any classifications you might be eligible for to reduce premiums. For example, different roofing materials (like sheet metal) require off-site manufacturing. If the contractor both manufactures and installs the material, he or she will fall under a different classification. As an owner-builder, if you have the ability to do this kind of work, this is true for you as well. The idea is to embrace every cost-saving opportunity you can.

Errors may occur at all stages of the build, and by the least or the most educated and experienced. For example:

CASE STUDY: PATRICK LISKA

Liska Construction Company

Verona, New Jersey

goofybuilder@comcast.net

"I remember one home we were building where the elevations for the house were taken right off of the construction plans. The house was built, a neighbor complained about the house being too high ... We checked the building elevations and they were correct; only thing was, the architect messed up on his elevations from grade. Needless to say, we ended up cutting the mansard roof off and lowering it while adding steel to help support it."

In this instance, the architect was at fault. He or she most probably carried errors and omissions coverage (E & O), as do most professionals. However, depending on the terms of the insurance, the insurance carrier may not have approved every expense. For an example, say the error caused an obvious delay and the family was required to pay rent for an extra three months for the apartment they had leased. As a consequence, the landlord may demand a new lease, sure to be broken. What if the family had to bear the expense of moving their furnishings into storage for that period? The questions go on and on.

The best solution is, of course, prevention. Speaking to a lawyer and insurance agent before the building process even begins is the best defense down the road.

We will discuss contracts at the end of this book. Suffice to say at this point is, unless you are a lawyer, you are unlikely to have the ability to draw a comprehensive contract that encompasses all the possibilities of contractor problems.

Speak to a lawyer well-versed in building and real property law, then to an insurance agent for the liability coverage you require.

CHAPTER 3

Planning and Scheduling

3.1: Planning a Timeline

Creating an accurate timeline for your build will depend on many factors, which could extend the construction period anywhere from six months to one year. What determines the actual time is often not under your control, such as weather conditions. The number of workers subcontracted and the amount of delays are also key factors that push timelines beyond the original plan. Understanding the difference between what you expect to happen and what will *actually* happen will play a key role in this portion of the planning.

As an example of a basic timeline:

- **Month One:** This is where the majority of your planning should take place. If one month of focused planning occurs, there is less risk of running into problems throughout the build. During this month, plan out every aspect of the build, right down to the amount of nails that will be needed. If you need longer than one month, take that extra time. Additional planning upfront does not cause budget issues.

- **Months two and three:**
  - o   Obtain all necessary permits
  - o   Complete all the ground work
  - o   Pour the foundation
  - o   Rough-in plumbing and electrical[1]
  - o   Frame and install sub-flooring

- **Months four and five:**
  - o   Frame up all the walls, including window frames, door frames, and the roof
  - o   Finish installing sub-flooring
  - o   Build exterior walls and sheath the roof
  - o   Frame interior walls
  - o   Finish roughing in electrical

- **Months six and seven:**
  - o   Finish off the roof
  - o   Install windows and doors (exterior)
  - o   Finish interior walls
  - o   Trim the exterior
  - o   Apply outer-wall finishes
  - o   Finish the plumbing and electrical

- **Months seven and eight:**
  - o   Install interior flooring
  - o   Hook up appliances
  - o   Final inspection

---

[1]The term "roughing-in" refers to a plumbing or electrical system that is completely installed and ready for fixtures.

Notes From the Field

Take advantage of every opportunity you can to speak to other owner-builders about their experiences with creating a timeline. These conversations are successful in-person, as well as in forums and chat rooms on Web sites. You will hear first-hand accounts of what did work, what did not work, and what the owner-builder would have done differently if given the chance. Take careful notes during these conversations because this information can prove to be valuable.

The timeline you create will be more detailed and complex, based on the amount of work going into your home, house size, and how many people are working on the building site daily. Setting up this timeline, though, allows for goal-making and the ability to stay on top of what is happening and when. Establishing this forecast can also allow for predictions in weather patterns. As an example, if the build starts in April, then you know that you have until September to get the construction weather-tight before snow starts falling, which is an issue in colder regions. Other weather conditions that may hamper your plans are monsoons, hurricanes, or tornados. Depending on where you build, you will have to plan around these naturally occurring inconveniences.

Be honest with yourself about the fact that delays can occur. In addition, demand honesty from your general contractor, subcontractor, and tradesman so you can be prepared for everything. For example, if they say it will take six weeks to complete a particular task, add an additional two-week pad. You will be surprised at how scheduled padding will be utilized and needed more often than not.

Money-Saving Tip:

Being as detailed as possible with your timeline will save you money throughout the build. Not only will taking the time to work out every kink allow for you to stay on schedule, but it will also prevent the need to make room in the timeline for things that have been forgotten. If the forgotten things affect the budget, you are losing money. If you need help with the details, involve your friends, family, and other owner-builders to help you brainstorm. Planning will cost you nothing, but not planning will.

3.2: What Types of Schedules Are There?

The next step is devising a workable schedule. This agenda must be realistic in order for this build to be successful. How many times have you scheduled something in your life, be it personal or professional, and an event arose that requiring cancellation or delay of a scheduled plan? This kind of interference is going to be a commonality in your owner-builder experience, as well.

It will not be uncommon to feel frustrated, let down, and as if you are not dealing with professionals. The first thing you must understand is that you have to expect these occurrences. Like you, these contractors are human; they have limitations, and they are dealing with other people who may set them back. But also understand that shipments will not always arrive as quickly as you would like, or be 100-percent accurate at all times.

You will want to set up both a monthly and a weekly schedule, allowing for mishaps, illnesses, and holidays. The monthly schedule should be an overview of the entire project, and the weekly schedule should be a more detailed outline of tasks. Each schedule should be completed prior to the building process. The weekly schedule will be heavily used, edited, and scrutinized throughout the build. These schedules, though not set in stone, should be accessible to everyone involved in the build, including members of your family. Make several copies as soon as the final draft is completed so they are on hand as needed to pass out.

Setting up the monthly schedule should be done in a general manner. This will serve as a basic guide that shows you what needs to be done on a weekly basis. This guide will be vital when the bidding process for subcontractors occurs because it will give you an idea of when you need different subcontractors. You will want to base your interviews on this schedule, so have plenty of copies to pass out at the time of your meetings. An example of a monthly schedule is located at the end of this book in Appendix C: Worksheets.

Using the notes created in your monthly schedule, you now have the ability to create a more detailed weekly schedule; this is the most crucial schedule. Be prepared to make a lot of changes to this document on a continuous basis because the weather conditions can make a bad turn, deliveries can be postponed, and subcontractors can become delayed — or finish early. Find a method of creating this schedule that works best for you, be it handwritten or through use of software. An example of a weekly schedule can be found at the end of this book in Appendix C: Worksheets.

This schedule should be kept with you at all times. Mark off items as they are completed, make notes about delays, and make any other necessary additions as they come up. Do not leave out any details because later when you are trying to solve a problem, this schedule might hold the answers you are looking for. Each time a change involving the contractors or suppliers is made, recreate or edit your final schedule to provide updated copies to everyone working on the building site. You will quickly learn how beneficial this small communication tool is.

Notes From the Field

Owner-builders who normally are not organized have reported success by using a system of breaking things down similar to what is outlined here. That which is challenging is not impossible. However, if you absolutely cannot seem to get yourself together when it comes to organizing all these details, you are not alone. You might have to ask for help.

Next on your list of planning and organizing is setting up a workable calendar derived from the schedule created. This calendar must be communicated to your subcontractors to keep your build on schedule and to keep every relevant person informed. By staying coordinated, all the management aspects of the build will run smoothly. Finding a calendar that works best for you will ensure productivity. For example, if you are a visual person, a large desk calendar is the best option.

Communication is going to be the biggest part of staying on schedule, so it is vital that your subcontractors can reach you at all times. Provide them a cell phone number, and if you are not available, be prepared to answer messages promptly. Daily planners are also effective organizational tools for those who prefer classic pen and paper.

Do not forget to create a list of items from your supply list that might take longer to special order or custom build. This includes windows, doors, and any specialized cabinetry. Communicate with manufacturers about what is considered typical stock and what takes time to come in. This information must be accounted for in all the schedules created during the construction process. Allow a couple weeks of cushion for these items to arrive in stock.

Be prepared for what is often referred to as the "domino effect." This occurs when one contractor hits a delay and causes the other contractors to be delayed. This is an inevitable reality because wrong material shipments come in, people become ill, and things break. Be flexible.

Money-Saving Tip:

A lot of the materials needed for planning and organizing can be purchased inexpensively at dollar stores, on clearance in big-box stores, and at surplus stores. You can find what you are looking for at these locations and you will save money. Do not be afraid to bargain-hunt for these items. You do not need anything that is top-of-the-line or fancy — but you do need things that are functional.

3.3: Why Should You Plan Accordingly?

The key to a successful owner-builder project is proper planning. Many owner-builders make the mistake of rushing to break ground, but this is not good for anyone. The previous chapters have outlined what you need to do to create a successful plan, and this chapter will address why. The last thing any owner-builder wants to go through is a project that does not run smoothly.

It costs nothing to plan, and it causes no one delay. Planning ahead allows for bargain hunting and the advantage of opportunities and considerations that normally do not present themselves without a proper plan in place. Do not gloss over any detail; if you do, you may suffer delays, budget problems, or other problems throughout the build. In addition, if you remain prepared at all times, you will be taken more seriously as an owner-builder. If you are viewed as being knowledgeable and level-headed, less time will be wasted discussing and debating details and decisions.

Keep the old adage in mind, "Measure twice and cut once." This is true for what you are doing right now. You are ensuring the proper cut (plan) is being made the first time so events will not go spiraling out of control. Remember you are in charge so be in charge. It is not uncommon to become worn out or lose interest during this process, but it is just as imperative as the actual build. You will experience both good and bad days.

Notes From the Field

We have never heard it reported that too much planning or being too meticulous was a waste of time — nor have owner-builders reported that it did not work to their benefit. They have reported, though, that failure to do so has cost them a great deal.

After you decide to hire a contractor, they will want to discuss scheduling with you regularly. This will ensure their crew is working efficiently and that the deadline will be met. They characteristically tell their crew when they are four weeks, three weeks, two weeks, and a few days away from being on your site. This notification system keeps everything in line, and when the crew knows what to expect, they know how quick and efficient they need to be. This also allows the contractors to line up future work, so staying on schedule is imperative to running their business.

If it sounds exhausting and like a lot of hard work, do not be dismayed. It is not an easy task, but planning is worth every effort. It is not impossible, but you need to know what to look for and where. Be sure you have an effective system in place that is well thought-out.

Discuss the components of your schedule and plans with your peers, your family, with other owner-builders, and with people who have had similar experiences. Listen to their advice and to their feedback. Stay objective, but do not dismiss any of the information they are providing for you. This information will prove useful more often than not.

Every minute of delay translates into loss of your money. This fact will be repeated numerous times throughout this book. If the project is delayed by several months due to improper planning, then everyone loses money. In some cases, tradesmen experiencing delays may apply additional fees to bills. Thus, it is vital to stay on top of planning and scheduling at all times.

Money-Saving Tip:
As much as you are going to hate doing it, you must add the possibility of firing non-performing subcontractors and tradesmen to your list of roles as an owner-builder. These people will cost you too much money in the long run. No one benefits from keeping someone who is not working effectively on the payroll. Be strong.

3.4: Act As If You Are a Business

You are going to be the planner and the organizer. Your responsibilities will include ordering supplies, giving direction to subcontractors, being the accountant, manager, secretary, custodian, researcher, and record keeper. All of this and more are the functions of a general contractor. Take note that while acting out this role, it is vital to also keep proper business practices in play. Being business-like during all your interactions projects leadership through the decision-making process.

Notes From the Field

Think about how you view a businessperson you respect. Mirror their business practices and work ethics while setting up your business persona. This will give you a good place to start and help establish a level of comfort. Establishing a "work personality" will also give you a basis of comparison for what works and what does not work when handling people in the business world. The implementation is going to be challenging, but that is par for the course during this entire building project. You will become used to being tested as time goes on.

Do your best to resemble others in the industry you look up to. Put a name with your project, get some business cards, and give yourself a title. You do not have to be incorporated to go through these motions. Not only will a business name and a title deliver some unexpected prestige, it will motivate you to work harder and smarter. This is particularly true if you are undertaking a large building project that will span a long period of time.

If you do not have a fax number, use the fax number available to the public at your local copy shop so that whenever a fax comes in for you, you can pick it up. Add this fax number to your contact information on all pieces of correspondence you send out and to your business cards. With the amount of paperwork, documents, and invoices you will be passing back and forth, having access to a fax machine will be a beneficial timesaver.

There may be advantages to you if you obtain a tax identification number — some states call it a sales tax number, and others call it a resale number. Many stores offer wholesale rates to holders of tax numbers, and they may not to others. In most jurisdictions, there is no fee associated with this identification number. Be sure to pay sales tax when purchases are made for use on your own property. If you have any questions or concerns at all related to taxes or tax identification numbers, contact a financial advisor, an accountant, or a certified tax-preparation specialist. It is better to do this than to make mistakes due to unanswered questions.

Obtain a business license, in some cases called a Doing Business As (DBA) license, from your local town or city hall, usually the clerk's office. Have the license numbers printed on your business card and any other correspondence sent out from your office. In some cases, you may not need to obtain a business license because all you need is a tax identification number to open a business account. It is best to verify first, however, to save time doing paperwork and scheduling meetings.

Each time the communication of instructions occur, submit it to the recipient in writing. Whenever this takes place, both you and the contractor should sign and date the correspondence. Keep copies for yourself in a log of correspondence so you can easily access the copies in addition to providing copies to the contractor. This practice will avoid any disputes about what or how things were instructed. This is a good example of how useful a fax machine is.

Fax machines, by the way, are a small investment. You can obtain a combination printer with copier and fax capabilities for less than $100, and a refurbished one for around half that amount. Compare that to the $1 or more per page fee from the copy store and the convenience of having it in your office, and you may well have a return on investment within the first two months.

Also, remember to speak as courteously as possible during all verbal interactions. Yes, things will go wrong; yes, things will be forgotten; yes, people will be frustrated. But handle all situations with a level head. Rather than acting adversarial, listen to and communicate concerns objectively. Not only will the contractors respect your position as owner-builder, but you will also respect them more. You do not want to get into a serious argument with your contractors. You need them to complete their work on schedule, and unhappy contractors have a way of finding reasons to slow down.

If you find yourself in a situation in which you are not respected, maintain your decorum at all times. Not only will your behavior prove this individual or group wrong, but it will also command the respect you deserve. Be strong; you know what you want to accomplish, and you know what you are doing.

Money-Saving Tip:

By performing all of the "business" tasks yourself, you are automatically saving money. Not only do you not have to pay someone else to perform these tasks, but you are your own best deal-finder. Your involvement also ensures that time is always being spent wisely, and every necessary task is being accomplished on time.

3.5: Planning Steps

To better plan your building project, break it into categories. These categories include:

1. **Building site search:** As reviewed previously in this book, this is a daunting task requiring plenty of time and patience.

2. **House plans:** This can be time-consuming if you are particular, but if you have a good idea of what you want, it will not be too bad.

3. **Financing:** Do not overextend yourself. This can happen quickly if you are not careful.

4. **Budget:** Follow this to the letter. Otherwise, your building project could be seriously delayed or — worse — completely fold.

5. **Permits:** Nothing can happen without all of these. Remember, you will be fined if you try skipping this part of the process.

By breaking the process up into categories, you can better understand the planning steps involved. Further break down each category based on the specifics of the build you are planning, where you are planning to live, and how much you are planning to spend. If you break down these categories even more extensively into smaller chunks of specifics, you will have a better understanding of how doable all of these tasks are.

Studies show it takes about six to 12 months to plan an efficient owner-builder construction project. These same studies have also proved that when the owner-builders short themselves during the planning process, they lose money and the build takes longer. As you can see, planning is extremely significant. The last thing you want to experience are delays or the costs associated with them.

When committing all plans and schedules to paper, you will finish your project smoother and faster. Proper planning is the key to your success. You will grow tired of this process, you will become worn out by this process, and you will feel burned out by this process. Pace yourself accordingly throughout it all. Enlist the help of a spouse, a family member, or a trusted friend to alleviate some of the stress.

Money-Saving Tip:

Following your plan will save you money. Do not forget, as mentioned previously, that time is money. Time wasted is money wasted. Follow the planning steps you commit to paper, and place these same expectations on any contractors you hire. It will be much easier to stay on budget this way. If you do not stick to the plan, it is going to be difficult to place that expectation on anyone else. Be disciplined and others will follow suit.

3.6: Essential Tools and Equipment

It is expedient to start looking for the tools that all homeowners need be-fore you actually need them. There are many name-brand and superior off-brands available through some of the national chains, but for the bigger and more specialized items, it would do you well to explore renting from a local vendor.

As a general contractor for your owner-builder project, you will need some tools that are necessary for completing your day-to-day tasks. The follow-ing items are essentials you will need to maintain optimum performance:

- **Large Dumpster:** Waste management companies rent these out for a monthly fee. Expect to pay additional fees for dropping it off, renting it per month, and dumping it out on a routine basis. Some management companies will roll this all into one monthly fee, which can bring the cost down for the consumer. Save money in the long run by choosing the largest size you can. Be sure the Dumpster is placed in close proximity to the work site for easy disposing of materials. It must be out of the way, though, so that heavy equipment can move in and out easily.

- **Portable toilet:** These can also be rented on a monthly basis. Your subcontractors will be grateful. Be sure it is not located in a muddy area, and have it maintained as often as the company recommends. Keep it clean to avoid contaminations and reduce the amount of sick days that could occur.

- **Portable electric generator:** Until temporary service is connected for power, this is going to be necessary for running power tools and shop vacuums. Some subcontractors come prepared with genera-tors, but you will find many that do not. It is best to be prepared.

Even if the contractors arrive with their own generator, you will find that having a back-up is useful, especially if you plan to work after hours when everyone else has gone home. These can be rented or purchased. If it fits your budget, consider purchasing one as a back-up system for your home.

- **6-foot ladder:** You will be using this ladder often, so be sure to purchase one that is sturdy and high-quality. Do not depend on the contractors for supplying ladders for you to use. If you are working on the building site after hours, you will not have access to their ladders when they pack up and leave at the end of the day.

- **16-foot extension ladder:** A lightweight ladder will work best for getting on the roof, painting, putting up siding on a two-story home, and when the house is complete and external maintenance is required.

- **Basic power tools:** Topping the list should be a good circular saw, an electric drill, a cordless drill, a set of high-quality drill bits, and a shop vacuum. You may wish to purchase back-up batteries if they do not come with the cordless sets already. Good quality may be obtained at reasonable prices if you check out the discount tool stores in your area and online.

- **Basic hand tools:** You will need a 25-foot tape measure, a hammer, pliers, screwdrivers, a framing square, a 6-inch level, a 4-foot level, shovels, and a sturdy toolbox. Find a comfortable tool belt because it will be worn daily.

- **Miscellaneous:** Start out with a steady supply of nails and screws, in addition to those that were part of the original supply order.

These little items get lost, bent, and dropped more often than anything else. It is better to have too much than to run out.

- **Light-duty truck:** You will save money by picking up your own supplies, rather than having them delivered. You should only have materials delivered if they are too big to pick up yourself. Purchase a used truck that already has scratches and dents in it so you do not have to worry about the materials or whatever else you are hauling, shifting around or damaging the truck.

Notes From the Field

Some contractors bring in their own Dumpsters and portable toilets, so inquire about this during the bidding process to see how it affects the bid and to see if you can reduce costs by providing these amenities yourself. If you can cut costs by supplying them, then take advantage of the opportunity.

As you go through the building process, you will find that you need other tools. This list, though, will give you the start you need to get going for a while. You will also receive plenty of advice from people about what should be on the job site at all times. While you are working, create a list of tools and other equipment you might need to purchase as the needs arise. This supply list will give you the ability to price shop for the best deals, borrow items, or barter for items.

Money-Saving Tip:

The tools you use will depend on what tasks you undertake. The only rule is to not assume you will be able to use someone else's equipment. Mark all of yours with etching or permanent marker and keep them in a separate area than the workmen's equipment. If you feel some investments are unnecessary, do not make them. Use your best judgment and you will do fine.

The answer lies in whether they had given oversight authority and responsibility to a general contractor or did it themselves. The bottom line is that tighter control of the building process will probably cause the project to come closer to the projected timeline, but even with daily monitoring, the build always has variables that may extend the life of the project.

Being prepared is the key to building your own home. Some items may be competed ahead of schedule, but more often they are on-time or late.

CASE STUDY: MELANIE NILLES

Author and owner-builder

Bismarck, ND 58501

www.melanienilles.com

"Our builder pushed back the completion date a couple times. Luckily, the management at our apartment let us stay a few days longer, but we had to push to get out of there and into the house. Our builder let us start moving things into the garage a few days before close because she knew we had to be out. We still had too much stuff to move in one day with our pickup.

The builder arranged everything, but the electrical inspectors did not come out until we lived in our home for almost three years!

I wish we had known more at the time and had lived in another house first so we would have known what to expect. It was a bad way to start out as a first-time homeowner."

CHAPTER 4
Financing Your Project

4.1: Types of Construction Loans

Getting a construction loan can be a difficult process. This is particularly true if you are choosing to build on raw or bare land. Often, the lender will not offer financing on undeveloped property unless you are an exceedingly valuable or reputable customer. The first step in the process is to understand the different types of loans available to you. Your lender will offer a list of what you qualify for based on the following factors:

- Good credit
- Sufficient income
- Equity in the property

There are three basic types of construction loans available:

1. **Single Close Construction Loans:** These loans comprise one fee, one application, one contract closing, and one fixed rate.

2. **Owner-Builder Construction Loans:** This loan option is similar to the single close construction loan, except it caters to those who want to do most of the work themselves. Many loan institutions will not have lending packages for owner-builders, so shopping around will be required.

3. **Dome Home Construction Loans:** These loans are incredibly difficult to secure because of their unique construction and how difficult is it to find comparable finished properties.

Notes From the Field

Each time you have a meeting with a financial institution about the loan options available, request written material about the information discussed; this way, you can take it with you, evaluate it completely, compare it to other information gathered from competitors, and make an educated decision. These financial decisions do not need to be made on the spot, so do not allow anyone to apply pressure in this direction.

When talking to the loan officers, you want to be sure they are familiar with the owner-builder construction loan process. If they are not, you should either request a referral from them to a colleague or do some research to find another loan officer who is knowledgeable. Avoid institutions that require a consultation fee; reputable establishments will help you from start to finish without any additional fees involved.

There are some key factors to take note of when searching for the right owner-builder construction loan. Remember, these types of loans allow you to be the general contractor. These factors include:

1. **The loan allows you to be the owner-builder:** If the loan requires a licensed general contractor to perform the work, then you have

ruled yourself out as an owner-builder, and you will not be allowed to perform these tasks on your own. This is overlooked by some and will cause a lot of problems down the line. Discuss your plans as owner-builder with the financial institute until they find a product that matches your needs.

2. **The loan allows you to find subcontractors:** If overlooked, this is another feature of the loan that will rule you out as an owner-builder. If you cannot hire whom you see as fit, you are not acting as the general contractor. There are construction loans in existence that have lists of subcontractors you are required to work from, but this is not the requirement of all loans. Read the fine print and discuss your plans thoroughly with the loan officer.

3. **The loan allows for minimal out-of-pocket costs:** Finding an owner-builder loan that does not limit the amount of draws (with construction loans, there is frequently a four to six draw limit) will prevent out-of-pocket costs throughout the duration of the project. Keep as much control with you as possible.

4. **The loan offers the owner-builder protection:** There are owner-builder loans available that do not require any payments until you move into the home. This type of loan will be the most beneficial to you in the long run. You will not only save money, but you will also be able to prepare a realistic and reliable post-construction budget.

Money-Saving Tip:

In order to keep percentage rates lowered, which results in more money in your pocket, be sure your credit score is good, and that you are in good standing with your accountants. Several online companies will provide you with your credit report for free. See **www.freecreditreport.com**, **www.experian.com**, **www.credit.com**.

4.2: How to Use Debt to Your Advantage

Without having the necessary cash on-hand when going through the construction process, the whole project can come to a halt. To make good financial decisions, it is vital to educate yourself through classes, workshops, and books such as this one. The more financial management education you have, the better you will be at making the tough decisions.

Notes From the Field

Be careful here because your credit score, in addition to your debt-to-income ratio, could stop you square in your tracks. Speak to a financial specialist about these realities. If you fail to do so, you are setting yourself up for a lot of disappointment. Credit is the first thing financial institutions look at when determining if you qualify for a loan or not.

It may sound odd that debt can be a good thing, but it is true. If your debt on balances is reduced by 50 percent, then you are creating good debt. If you leave old accounts open on your credit report, you are showing solid reputable history and creating good debt. Speak to a financial specialist or advisor about more ways to create good debt.

Money-Saving Tip:

Although you can obtain free credit reports online, be aware of any memberships or trial subscriptions that may be involved. You are responsible for remembering to cancel the subscriptions and memberships; if you fail to do so, fees associated with them will incur. Print out a copy of your credit report from each credit bureau so you can see exactly what you have, what needs to be fixed, and what debt you can use to your advantage.

4.3: Shop for the Right Terms

Shopping around for the right terms and loan conditions is essential. There is plenty of fine print and red tape involved in the process, so do not get discouraged if you do not find exactly what you are looking for after visiting with a handful of loan officers. It is common for loan officers to have no experience with owner-builder construction loans, so be sure you are doing your homework with each turn this process takes. Be prepared to consult with dozens of professionals before finding exactly what you are looking for and are financially qualified to do.

Shopping for interest rates is not necessary during this phase of the loan process, What you are going to be doing instead is shopping for the right features, terms, and conditions. If the fine print reads differently from what you are looking to achieve during the build, you are receiving a red flag to move on. Either ask the loan officer for a different product or go to a different institution. This is a prime example of why you must speak to and meet with numerous professionals.

As with the subcontractor bidding process, it is important to find more than one financial institution to fund your project. This means narrowing all of your meetings down to three good candidates to choose from. Look at the features of each institution's loan, line by line, to find the best product for your project. If you can only find one institution able to fund your project, be sure to go through every detail and iron everything out until it meets all your needs.

Perhaps one of the items to discuss briefly is the difference between a lender and broker. A direct lender employs his or her own staff to take applications and see the paperwork through the loan processing. A broker will take your paperwork and submit it to the lender who he or she believes will give you a loan. Brokers, while many are indeed trustworthy, do not always have the client's interest as their first priority — rather, they have commission in mind. You will probably pay an application fee at either source, so be prepared, but it might be better to shop around lenders rather than shop around brokers.

Notes From the Field

Be keenly aware of the many non-reputable mortgage companies popping up all over the place. If you feel suspicious about their business practices, their terms, or their services, contact the Better Business Bureau (BBB), and check the BBB Web site (www.bbb.org) to see if any consumers posted any information about them. Arm yourself with as much information as you can because they are, after all, handling one of the biggest investments you will make in your life. If you have questions about a lender's ethics, do not use him or her, no matter what is promised.

During the shopping-around process, you want to be taken seriously by lenders, so provide a well-prepared presentation to them during your meetings. You will find lenders will be fighting for your business rather than the other way around if you sound like you know what you want and how to obtain it.

Money-Saving Tip:

If possible, try securing an all-in-one loan. This type of loan is appealing because there is only one closing, it pays off the construction loan, and the mortgage loan kicks in automatically.

This saves a lot of time and a lot of fees. Speak to your lender to see if you qualify for this type of loan.

Sometimes the best advice on how to handle the surprises, financing, and anticipation of building your own home comes from someone who has gone through the process. Here, Kenneth Woulfe shares his insights.

CASE STUDY: KENNETH WOULFE

Owner-Builder

Green Lane, Pennsylvania

As time passes, we often have a more objective, retrospective view of what has transpired. Those owner-builders who are happy with their completed project generally had good advice about planning, planning, and more planning. Those with poor experiences echoed that sentiment and were rather emphatic about knowing who you are dealing with. As one owner-builder shouted at the end of her interview, "What did we know when we started? We were stupid!" We asked Kenneth Woulfe to share his thoughts and experiences about the house he built several years ago.

"When starting any major project, planning is the most important step. It always seems to be the least rewarding when you are doing it, but, when the project is completed, it will easily be seen that the complete planning of the project was the one singular step that helped all of the other steps fall into place and have the project come to a successful conclusion. Keep in mind that a basket full of scrap paper before you begin is better than days and weeks of wasted effort once you have begun.

Building a home is the most challenging project that most average people will ever undertake. A complete knowledge of all facets of the building process is not necessary, but a good overall awareness of what needs to happen is certainly recommended. There are many good books that help the novice approach the home-building process, and it would be a good idea to reference several of these prior to starting the process.

Most anyone who is going to tackle the job of building their own home would most probably have a good idea of the home they are going to build. It may be a dream home that has been stirring for a long while, or it may be a sudden change in life priorities that presents the individual with the chance to move forward into the building process. Many times the property available to the individual will actually dictate the type or style of home they are going to build.

Other times, individuals may have a home in mind, which they then use to search for the appropriate building site. In either case, it is very important to keep these two aspects in mind. The right house may be completely wrong if the setting in which it is placed is not appropriate.

CASE STUDY: KENNETH WOULFE

After you have the alignment of home and location decided, the average person acting as an owner-builder would be wise to find a good primary contractor to assist in the timing of the building process. While the owner-builder may participate with a little or a great deal of work equity in the process, the primary contractor will be the one responsible for making sure the proper steps are completed in the appropriate order. The primary contractor will also have access to many subcontractors that may not be accessible to the owner-builder. These subcontractors may well respond in a timely manner to a primary contractor with whom they have worked previously, rather than an owner-builder with whom they may never work again.

These have been some of the most important preliminary steps that I can think of to building your own home. Match the right home to the correct building site; obtain as many good reference books as you can and do your homework to keep you on track; decide ahead of time just how much work equity you are willing, able, or have time to contribute; and, finally, find a good contractor with whom you can work and who has the proper references and credentials."

CHAPTER 5
Finding a Building Lot

5.1 Location and Geography

The type of home you want to build will often be dictated by your geography. An adobe home in Washington state is probably not reasonable given the high amount of rain and moisture, which will rapidly deteriorate the dried mud bricks. Conversely, building a log home in the desert where termites, fire ants, and other wood-eating insects run rampant is also not smart. The type of ground will determine whether your planned house will work without blasting. Also, there are not many homes with basements in the desert or parts of Alaska. Know where you are constructing and what is feasible.

Be wary of land that appears to be a bargain. Find out if it is in a floodplain and whether previous potential buyers turned it down. Know your neighborhood, as well. By law, real estate salespeople are not permitted to inform you of certain things about the property. If it matters to you whether someone died or was murdered on the property, if the surroundings are allegedly haunted, or even whether sexual predators live in the area, then you should talk to the real estate editor of the local newspaper, ask neighbors what they know, and then ask the owner again. If the owner does not tell you what you have already learned from other sources, you are not dealing with an honest individual, and it may not be in your best interest to purchase the property — not because you believe in ghosts, but because if they are deceptive with one thing, they may be deceiving you on something far more important. If you are concerned about finding sexual predators, you can check with the local police or go online to one of the many Web sites that will provide such information. Each state has their own database, usually through the state police or local law enforcement. There is a national database you can access at **www.nsopw.gov/Core/Conditions.aspx?AspxAuto DetectCookieSupport=1**. Conducting this search is imperative, especially if you have children.

Yet there are other considerations to keep in mind when finding a site for your home. Whether purchasing a "fixer-upper" or starting anew, keep in mind the old real-estate axiom: "Location, location, location."

Should you seek to build a full or partial solar home, the panels and windows will need to face south because that is the most consistent source of sunlight. Is there enough sunlight from that direction? Will trees need to be pruned and cleared, or is there a structure blocking the southern view? When selecting a location, these should be primary questions for a solar-powered home.

Building a swimming pond or pool brings more significant deliberation when looking at land. It needs to be checked to see if you are on top of bedrock, or if the ground is suitable to support your structure. Equally important is how high the water table is. A high-water table is good if you need to drill a well or want a natural pond, but not so great if you plan a deep foundation, as with a basement.

While we will discuss zoning shortly, think about any special hobbies or work you do. If you restore cars, you will need to see if there is adequate space to build extra storage. A radio operator may run into neighborhood opposition when if radio tower goes up, even if zoning allows for it. Rural homeowners can be just as passionate about their property as those who live in urban areas.

Budget wise, rural areas are less costly than those closer to metropolitan regions. If you do not mind commuting, you may find affordable land and lower construction costs away from the city. The flip side to that is the commuting time, tolls, parking, rail fees, and inability to run home for lunch. It is a trade-off, and you will have to decide which is more important and makes most sense.

Considerations: Are fire and ambulance services available? If not, where do they come from and what is their response time? This is problematic for families with small children, for the elderly, and for those with medical conditions.

Some people choose to live miles from their nearest neighbor, while others cannot exist without the noise of neighbors on the streets at all hours. Whatever your preference, there are some factors to ponder when finding the ideal location for you.

Be a Detective

Prior to moving into your neighborhood, you are going to want to research the following list of questions:

1. Where is the nearest grocery store?
2. Are you close enough to your current bank, or do you need to transfer your accounts to a new one?
3. If you attend church, where is the closest one to your new home?

4. Do you know where the post office is?
5. Where is the car wash?
6. Do you know where you can get gas?
7. Do you know where to order pizza or other food for your future workers?

These are the kinds of questions that must be answered before you move into your new neighborhood.

5.2: Finding the Perfect Building Lot

Choosing the perfect building lot is just as important as deciding what kind of house to put there. When it comes to finding that perfect place to build your castle, the first instinct is to find a real estate agent. Though some agents may have negative reputations, the industry is highly regulated, and the horror stories of years gone by are rapidly diminishing. There was a time when agents were notorious for getting the highest price possible because their commission depended on it, even if they were representing the buyer. Today, however, there is a strict code of conduct and ethics. Yes, the buyer can find property on their own by reading the ads, traveling to neighborhoods, and researching the Internet. Most Realtors® have lots and land to show, as well as other brokers in the area. It is easy to say not to bother with an agent, but there are two realities many instructional guides fail to discuss:

1. The real estate salesperson generally knows the area and nuances of the neighborhood, including the people.

2. Any property a broker lists makes the seller contractually obligated to pay the commission no matter who is responsible for the sale, with the exception of a very few states.

By all means, try to find property on your own — just do not be surprised if the majority are listed by real estate agents.

Create a list of questions to present to your real estate agent or broker to help weed out building spots, such as the following:

- Is it too hard to develop? Does it have a lot of ledge that needs to be blasted? Does it have too many slopes, or is it too wet?

- Is there thick tree growth or brush that would be difficult or expensive to clear and haul away?

- Is there a threat of flooding? There are solutions to this problem, but it is up to you whether you want to deal with them — at a price — or move on to a different piece of land.

- If it is an extremely rural location, is it too difficult to run a main electrical supply to? There are many pieces of land with high values but with no main electrical supply running to them.

- Can a well be dug if there is no town water supply? Be aware that if the property has a lot of ledge, it may not be possible to dig a proper well.

- Can a septic tank be installed if there is no town sewer to hook into? There is nothing wrong with a septic tank, but be prepared for the cost of installation and related maintenance involved.

- Are there too many size limitations set by a planning board or homeowner's authority? In some communities, these building rules and codes are in place, and it is your responsibility to find out about them.

- Has the land ever been contaminated? Obtain a copy of a natural hazard disclosure to see if anything has happened on the land that would contaminate the soil.

- Is the location too noisy? This is not limited to city sounds. You might be building next to a farm with a herd of loud cows or a choir of pigs.

- Is the shape of the property too odd? Look at the plot plan for the land to determine whether the property lines are too complicated.

- Is there a reason that it has never been built on before? Sometimes, local residents might have information about this situation, so it might not hurt to do some detective work.

Notes From the Field

Do not settle for a building site just because it is in your current budget. Over time, you may be faced with costly issues you will not be able to afford to address. Find the right property for the right price — or wait. The trick is to focus only on locations that you can afford, and explore as many as possible. This reality is difficult for the impulse buyer who takes the first building site that interests them and has to build the home right there. This is a mistake. In every instance, do your homework.

In addition to your list of what you do not want, compile a list of the aspects you might like to see on your building site:

- Is it zoned for residential use?
- Are utilities present?
- How far does the house have to be set back?

- Is there access to the lot?
- Are there drainage issues?
- Are there slope issues?
- Is the title clear?
- Are there any liens or other disputes on the property?
- How many neighbors are there? How close are they?
- What is proposed for future construction of roads and commercial structures?

These lists will prevent time wasting when it comes to setting up appointments to visit building sites. A list will also ensure you are getting closer to finding what you want. Add or remove items from these lists as needed so they fit exactly what you are and are not seeking. There is no such thing as asking too many questions. If you do not receive clear answers to your questions about the property, it may be a red flag to move on and inquire about another location.

Be sure to obtain the most land you can afford to prevent outgrowing the property. Long-term living is the goal for those building their own home, unless they are career builders who build and move on later. Focus on what the land will be used for, what kind of landscaping will be occurring, and how many other structures (like a garage, sheds, or patios) will be on the property. Failure to look at all these possibilities, as well as additional building situations not listed here, will cause a lot of problems and frustrations in the future. If you feel for any reason at all that the property is not suitable, do not settle just because your spouse, real estate agent, or the seller is just trying to make a sale.

Do not forget to consider any drawbacks the land might have in terms of community or region. Research if there are any dumping sites in the area, which devalue the property and can pose a health risk. Doing specific research in areas such as these will not only prevent unexpected and unwanted surprises, but will also bring some peace of mind during this tedious process. Meet the neighbors before signing your name on the dotted line, especially if you are in close proximity to each other. You would be surprised by how a little kindness can garner good will and even better information. Can you make friends with them, or can you foresee constant arguments over boundaries, noise, kids, cars, or music?

It is also important to choose land with a high grade. This will help ensure there will be no drainage problems. The last thing any new homeowner wants to deal with is a leaky basement because he or she chose a piece of flat land with poor drainage. Understandably, this is unavoidable sometimes, but those who have a choice should be exceptionally aware of the land's grade, how it slopes, and where the house will be positioned among these features.

If you are unable to find land with beneficial sloping, and you do make the decision to make a purchase, speak to your excavation contractor about how to solve this problem. You may want to be aware of the costs associated with solving any problems that might arise from the issues associated with the land. These issues could quickly put your budget at risk.

Money-Saving Tip:
Use town halls, city halls, and information centers to the best of your advantage. These places provide a great deal of information about different parcels of land, as well as what type of neighborhood they are located in. Collect as much data as you can in the way of conversations and free literature. Take careful notes if printed materials are not available.

Decision Matrix: Draw a list with three columns. In the first column, list the property with a name you will identify with. The second column is where you will list all the good features of the land. In the third column, list the negative aspects of the property. Stand back and see which properties have the most pluses and least minuses. You are now ready to make a decision.

5.3: What is the Difference Between a Lot and Land?

Finding a building site can be difficult, depending on where you are looking, so it is important to understand what you are buying. A lot and a piece of land are two different things. A lot is property that is ready to be built on, and it is normally one or two steps away from being ready for a utility hookup. Land is a word normally used to describe property that is unfinished and needs to be developed.

It is frequently more desirable to buy a lot in order to keep land preparation costs down. It is also easier to get a loan for a lot, in comparison to land, with a lower down payment. Lending institutions favor property that is already developed and has accessible utilities. Raw land — land that has not been developed at all and is not ready for building — in most cases can be considerably more expensive, so keep that in mind when searching. As a result, this land is less marketable and more difficult to secure a loan for. If, on the other hand, you do decide to purchase a piece of raw land, calculate the extra time it will take for clearing, building roads, extending utilities, and getting permits into your timetable. This can increase your schedule by months or, sometimes, years.

Money-Saving Tip:

Real estate agencies often have free information available on-line about different properties, including land and building lots. Print this information out and get as many specifications as you can. You may be able to meet directly with the real estate agent or the property's seller to obtain more information to influence your buying decision. If the seller is unwilling to get into detail about the land's specifications, move on.

5.4: What is the Value of Your Building Site?

There are several factors to consider when determining the value of your building site. An appraiser will look at other building sites in the area comparable to the one you are interested in purchasing to come up with a figure to provide the lender. Bear in mind this figure is not necessarily the resell value. In some cases, the appraiser will give you examples of the building sites comparable to yours to give you a better understanding of how they arrived at their evaluation. Be aware, though, some appraisers are busy and may not have the time to provide you with this information.

The difference between the appraised value of a property and the market value is simple. A certified appraiser inspects the property and determines its estimated appraised value or worth; banks and lending institution always do this when applying for mortgages. An appraiser arrives at their determination through comparison of sales, the condition of the property, proximity to transportation, utilities, zoning, and more. The market value is what the property will sell for. Every feature is dictated by the strength of the market and the demand for similar properties on the day of the sale. Your property can fall above or below the market value on any given day.

Several years ago, a friend was living in an upscale community with two vacant lots: one in front of hers and one on the side. Across the road from the front property stood a fenced-in stable and corral of a neighbor. The builder could not sell the land for more than three years and blamed the horses' presence. His asking price was $98,000. That was the market value before, but no one bothered getting an appraisal because it was not financed. This friend was too anxious to have the property and jumped at the opportunity, managing to whittle the figure to $95,000. Had this deal been financed, the friend would have learned the true value of the land was approximately $10,000 below what was paid. An appraisal before closing would have cost a few hundred dollars but saved her much more. This is a buyer-beware for those making a purchase without a bank or lending institution.

It should be mentioned that this friend used a real estate agent with less-than-admirable scruples because she was unaware of the requirements brokers are obligated to comply with. Their first sworn duty is to protect the interests of their client. This agent did not work in her client's interests, but rather encouraged the client to the higher purchase price of $440,000 for her home so that her commission was higher. You must always be conscious that there are unprincipled individuals in every profession. If you are not comfortable with a price, then walk away and tell your broker why. The moral of the story is that you should be confident the broker is working in your best interest.

Notes From the Field

For the best hands-on lesson about appraisals, request to be on-site when it takes place. Do not be afraid to ask questions throughout the process, and request any written material that might be available. Do not be put off if the appraiser does not have time for you to tag along. Remember, they are just as busy as the code enforcement agents. Because their time is spread so thin, you may not be able to explore this option. It does not hurt to inquire about it, though, so do not hesitate to do so.

The amenities available on the building site will play a large role in determining the value. Do you need to install a septic tank? Does a well need to be dug, or does the water need to be piped in from a town supply? Look closely at the amenities available prior to buying the land. When researching the value of the land you intend to purchase, ask the seller or real estate agent the following:

- Does a driveway (gravel, asphalt, or concrete) need to be installed?

- Is the town or the homeowner required to install the sidewalks?

- Does the town provide street lighting, or is the homeowner responsible for it?

- Does a lot of groundwork need to be completed?

- Does a special foundation need to be installed?

- Is electricity available to this site, or does it need to be brought in?

- How much will it cost to connect to the sewer if a septic tank is not needed?

- How much will it cost to connect to the water if a well is not needed?

Answering these questions will put you closer to having exactly what you want when you have finished your project. Remember, there are no stupid questions. If you cannot find the answer you are looking for from the seller or real estate agent, look for answers from the town, the city, the Internet, or even the subcontractors you will be working with. Chances are someone has the answer or the experience to solve the problem you have encountered. You might face a new set of problems by giving up.

Do not let this phase of the process frustrate you. There are a number of variables to come, including location, supply and demand, and the building site's characteristics. Speaking to knowledgeable professionals will help you effectively sort through this information. Where your home is going to be is just as important as how it is going to be built. This is just the beginning of the series of decisions you will be making.

Money-Saving Tip:

Brokers may or may not be an option. Understand that although the seller is responsible for paying the commission (average of 6 percent), the buyer really pays for it through increased cost built into the price.

If you buy through a broker, make sure that broker is working for you and not the seller of the property. The commission will be split, but you will have someone in your corner who is, by law, looking out for your interests. That broker is responsible for doing most of the leg work, finding out answers, and warning you of known hazards. Using a broker can be advantageous, but it is also more costly. You have to determine if you can accomplish the same thing they do, have the same access, or have the time doing research.

5.5: Should You Deal With Tearing Down Property?

Buying a piece of land or a building lot are not your only options as an owner-builder. You also have the opportunity of choosing a piece of land with a building unfit for habitation[1] on it, one that can be added to or significantly fixed up. Not only will this give you the experience of working with a building project, it will also offer you the opportunity of moving into an established community. In addition, you will have the benefit of all the utilities and other amenities present.

---

[1]Unfit for habitation is a dwelling that is unsuitable for people to occupy, meaning it may lack structural integrity, walls and or floors are weak, decaying, have holes, etc. The building may have a sound structure but lack proper ventilation, heating, plumbing. It may be rodent-infested or lack sufficient fire egress.

Notes From the Field

When doing renovation work, there can be extensive and time-consuming demands. For most, the effort has been well-worth it. Though some stated it took up to three years to complete, it was not with regret. Others have purchased pre-existing buildings only to tear them down because they liked the location. Each reported a sense of fulfillment and gratification with the final product.

Demolition, conversely, will prove to be a costly addition to your budget. Find three demolition contractors to obtain bids regarding the demolition and clearing of the property before making the decision to work with this land. What you save on utilities might not cover what it will cost to get rid of the old house, so be aware of that factor. Other factors to consider are:

- The property might contain hazardous waste, such as asbestos, that professionals need to remove. Do not attempt to remove this material on your own, as it poses a health and environmental hazard.

- If you leave one or two of the walls up, you might receive a tax break for doing a remodel as opposed to starting fresh Do your homework to find out whether the community supports such options.

- Most established communities require approval for major remodels. This is particularly true for those communities wishing to maintain the original curb-side aesthetics.

There are many benefits to working with tear-down property, which goes beyond the impact of the homeowner. For example, the locale might benefit from the added taxes collected from the building of a new home, or the neighborhood may have disliked the eyesore that was there before.

Take note that working with a tear-down property might not be favorable for the rest of the neighborhood for the following reasons:

1. They may believe the character of the neighborhood is being destroyed.

2. They may complain about how tall or how close the structure is to the property line.

3. Neighbors might not like the new design in comparison to the tear-down.

4. There might be complaints about construction noise and disruptions.

5. There is the chance of the rest of the neighborhood's taxes being reassessed after the tear-down property is reconstructed.

Researching the neighborhood prior to deciding whether the demolition makes sense is an important step for a couple reasons. The No. 1 reason is marketability. If you are building a house that is significantly larger or smaller than those in the rest of the neighborhood, it is not going to fit in, and the resell value will plummet. However, if you find a community containing existing or proposed homes similar to yours, this is the right direction to go.

Real estate agents tend to advise using tear-down property as a last resort. It makes the most sense to use a tear-down property when:

1. The location is a must. If the land or lot is exactly what you consider to be "prime real estate," you would be making a mistake passing it up.

2. It makes good financial sense. For example, tear-down properties tend to lower the land's value. Therefore, you are able to make a purchase at a better price and increase your building budget.

3. The buyer is willing to work with local guidelines. There are some communities and neighborhoods wishing to maintain aesthetics. If you like how the rest of the community or neighborhood is built, then you will fit right in.

4. There is a tight building deadline. For example, the foundation could be perfectly fine, but the building is not. Owner-builders will be saving a considerable amount of time on land preparation and constructing a foundation.

5. The emotional impact on the neighborhood is not an issue. Sometimes, there are factors involved with tear-down property, including historical preservation. Do your research to respect these conditions and considerations.

Money-Saving Tip:

Always follow the advice of a real estate agent when making the decision to buy tear-down property; they have the best first-hand information. While making a sale is important to them, it is unethical for them to point consumers in the wrong direction, and that would harm their reputation. The last thing you want to experience is dealing with a piece of land that will end up costing you more than building something new.

5.6 Buying Your Building Site

Buying your land can be an extremely confusing process. Before buying your building site, there are some important points to check up on. It is not uncommon to feel overwhelmed, anxious, and excited during this phase of the process because this is, after all, where you are going to be living for many years to come. Remember, there are pros and cons in every buying decision. Take into consideration the following factors when buying your building site:

What is the slope of the land?

While a flat lot is more favorable in terms of building costs and use, it might not always be feasible in the area you want to build. If the slope of the land is considerable, there are many things that need to happen during the preparation of the site and also to the structure of the house.

A house built on a slope costs a lot more to build than on flat land. Conversely, there are ways to make this a workable situation; an example is to create a walkout basement, which provides egress to the outdoors. This will allow homeowners to take advantage of the land, the space, and the views they are after. Be sure the plans account for this feature because structural considerations must be made and proper permits must be obtained.

It is also consequential to note that extra backfill, or gravel, is necessary in the foundation when building on sloped property. Enlist the help of a design professional and research house plans specifically designed for sloped properties. This will prevent unforeseen problems, as well as unsafe building conditions, from occurring. Speaking to a design professional will alleviate any concerns you may have regarding structural integrity and longevity of the building as a whole.

Where is your land positioned?

You will want a south side positioning so the sunlight will stream through the windows of your primary living spaces. If the land you have chosen is on the north side, this is still a workable situation. You may have the design plans flipped or a mirror image created. This is not a complicated process, so do not worry if this situation is entered into the equation. Remember, your home should be exactly what you want it to be, no questions asked.

Proper orientation of doors and windows in your home, while taking full advantage of the sunlight, conserves energy. Houses are meant to keep the heat in or out depending on the general climate of your area, so this positioning is vital for keeping costs down. Take this into consideration when developing your floor plan, otherwise you will not benefit from how the land is oriented.

What is the soil like?

Soil type has a big impact on the construction costs. No one knows for sure what is under there until the backhoe starts digging, so it is a good idea to prepare a reserve fund in case some surprises are uncovered. You can get a good idea of what the property is like by contacting your local building department. If you are not receiving the answers you are after from the building department, contact excavators and builders with experience working in the area. If all else fails, you may nave no choice but hold your breath and wait to see what the excavator finds when they start digging.

Some house plan designers create plans that are traditional to the area. For example, a house plan in New Mexico might call for a foundation that is significantly different from one in New Hampshire. Because soil reacts differently to water, it is significant to know what kind of soil you are working with and what kind of house plan is best suited. This is a key step in the land-buying process, so it cannot be skipped. If possible, speak directly with the house plan designer about any concerns you may have.

Is there private or public sewer?

Municipal sewage systems can be found positioned at the front of the lot, giving your groundwork crew a good indication of the level of the basement line. It all has to flow downhill, and if that is not possible, you may have to put in a grinding system or a pump station. These issues will have to be addressed in your budget. If the added cost of a pump is out of your budget, this building site is not an option for you to consider.

If there is no municipal access available, you will have to invest in a private septic system. There are several types ranging from the popular septic tank with leech fields to mound systems to aeration systems. You have to check with the local building department to see what the health district's regulations are because not all systems are approved. Again, the cost of adding in a private septic system is something to address in your budget.

Is it zoned?

Check with local authorities to be sure the land is zoned for the kind of home you are planning to build. While you are at it, ask about future changes in zoning, which includes retail development and any construction that might devalue the land. By failing to inquire about these details, some owner-builders have reported situations such as highways being built in close proximity to their new home. This situation could have easily been avoided by asking some simple questions. It may be difficult to reach someone right away, so be patient and be sure to leave messages.

What are its amenities?

Is it in close proximity to schools, hospitals, shopping, and the highway? While these factors affect the value of the land, they are also important to consider when determining how close (or far away) you want to be from everything necessary for day-to-day life. Do you want convenience? Do you want to be in the middle of nowhere? Calculate distances that work for you and whoever else will be occupying the home.

Do you need to prepare?

Land preparation can be costly in terms of tree removal, hooking up utilities, or dealing with large rocks. It is consequential to know what there is (a lot of ledge? a swamp? thick tree growth?) so your budget can be addressed accordingly. Despite how much you love the location, if there is too much work that needs to occur on the property, it can quickly drive you over budget. Though you and some helpful volunteers can do a lot, you have to be realistic about how much is too much.

Do you need to call in a professional?

During the buying phase, many owner-builders opt to use a real estate agent to get through all the legalese, understand how land is priced comparatively in the area, and understand the market for finished houses in the area. On the other hand, if you are interested in foregoing a real estate agent, information can be obtained through the title company. You can also research material on the Internet and from the local city hall or town hall. Utilize every resource you can to the best of your abilities.

Another pro to using a real estate agent when buying the building site is the preparation of an offer. This must be completed, either by the real estate agent or the owner-builder, to present the seller with a deposit check. This deposit is on average between 1 and 3 percent of the land's purchase price and is held in escrow until closing. At this point, if there are others interested in purchasing the property, negotiations will begin. Even if you decide not to use a real estate agent, one can still be consulted for advice.

Notes From the Field

Owner-builders choosing to use a real estate agent because the search process was too overwhelming and confusing have reported very positive experiences with their realtor. These same owner-builders have reported an increase in their ability to construct well-rounded building plans, as well as a more workable timeline. These factors should be weighed from person to person, though, and not based solely on the experiences of others.

Money-Saving Tip:

To offset land costs, the property should appreciate at least 20 percent each year. These land costs include taxes and interest (which are normally around 6 percent). Speak to a financial specialist about this information, and be sure you receive every little detail in writing, even if you have to write down what they are saying during the meeting or phone call.

Finding the best place to build involves a lot of leg work and quite frequently good luck. Such was the case when Patricia Baker was looking for land.

CASE STUDY: PATRICIA BAKER

Frisco, TX 75034

www.patriciasbaker.com

The land was for sale due to the commercial owner-builder deciding not to build a home on-spec because he had planned to start another development on a new site in the town north of their location. Ordinarily, businessmen prefer to sell the land and require any building be done by their organization. The builder was glad to let the new lot owners select whomever they wanted to do the construction. With a family friend in the building business, Patricia and her husband had the best of both worlds with a great lot and trustworthy builder who would accommodate their needs.

The Bakers were discriminating and, in the end, they found the exact property to suit their wants and needs.

CHAPTER 6
House Plans

6.1: Designing Your Home

You most likely have perused dozens of home design catalogs, books, and Web sites by now, and you may have even started sketching ideas out in your dream home notebook. It is an especially good idea to determine how the living spaces will be used in the home. The utilization of these rooms has a tremendous impact on how the house plans should be drawn. Use magazine cutouts, diagrams, and notes to brainstorm all of your information with your partner and friends.

Do some window-shopping by taking a ride through different neighborhoods with homes you like. Make note of the characteristics and features that you like, and also about what you do not like. Do not be afraid to take pictures of architectural and design features of interest to you. Take the opportunity, if possible, to speak to homeowners about their design and if they would make changes.

Make a list of the various features your home must contain. Keep in mind the size of your family, if you are planning for more children, and the type of entertaining you do. Here are some questions you should consider:

- How many bedrooms should the house have? *

- How many bathrooms should the house have? A practical consideration for resale is the more bathrooms, the more enticing the sale for large families.

- How many bedrooms should have a private bathroom?

- Should there be a separate family room in addition to the living room? *

- Do you prefer a separate formal dining room, or do you prefer it incorporated into the kitchen? It is also fine to have it as part of your "wish list" without a definite reason.

- How many porches and decks should there be? The more time you spend outside, the more outdoors features you will want to have.

- What size should the garage be? Fuel this decision by how many vehicles you currently have, if you have children who will eventually have their own vehicles, and if you have a need for storage. Do you want a shed attached to a garage or separate?

This list could go on further, based on your tastes and your family's needs. Once this list is complete, break it down further by each room. This mechanism will help guide the size of each room and give an indication of what spacing requirements will need to be accounted for. If you cannot stay within budget on the list, then start eliminating items that can be added later. Additions and expansions can always be made in the future if that space is not absolutely needed right now.

* Is your family still growing; do you have many guests?

Should you find you cannot build your dream house exactly as you want it from the start, consider a design that allows for these expansions. This will permit you to phase out your project as funds become available. A methodical plan will also allow you to figure out what is most significant to you at the beginning of the project, and then work your way toward the lesser important elements. While our first temptation is always to get what we want right when we want it, instant gratification is a recipe for disaster when making costly decisions about building a home.

How the house is designed and planned out also affects the resale value. While you may not want to think about selling your house before the building process even begins, it is an important item to ponder. Get advice from design experts and real estate agents in the area. They will have information about the local market and are able to give you ideas regarding how to get the most bang for your buck.

If you plan on buying a set of plans rather than drawing some up on your own, check out the Internet, home stores, or a contractor. There are numerous sources for obtaining plans, many of which can be customized to fit the needs and desires of each homeowner. Be sure to choose a reputable source, though, because some well-meaning contractors can be extremely pricey.

Notes From the Field

You will be tempted to purchase every set of plans that interest or inspire you. This can quickly get out of control and wind up costing a small fortune. Narrow down what you want before conducting any kind of search. This is especially important to keep in mind if you are an impulse buyer.

There are a few points to consider when choosing house plans. You will be facing both local and national codes, such as Uniform Building Code (UBC), Council of American Building Officials (CABO), or Institute of Building Control (IBC). To say these codes are complicated is an understatement. Most house builders choose to buy their plans because professionals who must know this information design these homes to fit the codes. When altering plans to fit your needs or fit the codes of your local code enforcement, it is imperative to make sure they still fit the national codes.

There are options that can help with this tedious process. Numerous Web sites exist in which you can enter specific information about the kind of house you want, how big the house will be, how many rooms you want, or the number of bathrooms you desire. Simply input the information and click "submit," and the site prepares a plan that includes the proper coding and other significant specifications. Remember, do not add anything to the house plan that is not absolutely necessary. Check out: **www.coolhouseplans.com**, **www.eplans.com** and **www.slhouseplans.com**.

Money-Saving Tip:

Consider doing a green build or adding green building elements to your design. Not only will these choices be ecologically friendly, but they will also reduce energy bills. There are many classes available, both face-to-face and online, to teach builders and owner-builders how to build green. You may find that these kinds of green building alternatives will increase the value of your home due to green building's rise in demand in the marketplace.

6.2: Determining Your Requirements

When determining your requirements, it is important to understand what is required of the house plans as well. Here is a list of plans required when building a home:

- **Detail sheet:** The architect regularly creates these sheets as a way of showing what materials need to be used and what needs to be done in each room. It is not uncommon to need more than one detail sheet.

- **Elevations:** These are drawings, normally numbering about four, showing all the compass point angles and the various views of the house from the outside.

- **Floor plan:** This is a scaled building and architectural set of drawings that show the rooms and spaces one floor at a time.

- **Foundation plan:** This is an above view of the foundation walls and footings.

- **Plot plan:** This shows what the property looks like and all the plans for the property, including what is going to be built and what already exists. It must be drawn to scale, legible, and easily understood. Significant features, such as where the septic tank will go and where the well is going to be dug (if these features are part of the property), must be clear.

- **Specification sheet:** This is an all-inclusive list of the materials — including appliances — going into the home that is compiled (sometimes with help) by the general contractor.

One of the big steps in the process is learning about the building codes for your area. They will determine what you can and cannot do, how tall your building may be, what materials are mandated for certain sections, and everything needed to build a house suitable to obtain a certificate of occupancy. You can obtain a list of local codes at your town office or city hall from the code enforcement agent or office. Below is a list of all the national model codes that must be followed when building a house:

- The Building Officials & Code Administrators International Inc. (BOCA)
- The International Conference of Building Officials (ICBO)
- The Southern Building Code Congress International Inc. (SB-CCI)
- The Council of American Building Officials (CABO)

Contrary to popular belief, there are codes for just about everything from how far the toilet should be set from the counter top to how many wires can be run through a hole in a floor joist. There are many ways to figure out which codes need to be followed. First, work closely with your jurisdiction and have open communication with building inspectors during the design phase. Then, be prepared to meet with inspectors during the build and take educational classes about the topic. Working with reputable contractors, also, will ensure proper code enforcement and save significant headaches.

Give yourself plenty of lead-time with all matters involving the code enforcement office. It is usually one of the busiest offices year-round. This means it will take several days for a returned phone call; it may require several weeks before a meeting can be arranged; and it could take months for an inspection to take place. In some towns, nothing proceeds without the approval of the zoning or planning board, which may meet monthly. You, or a representative, may have to attend the public meeting. If you are prepared for these facts, you will run into less problems and experience less stress. If you do not prepare yourself with enough lead-time, you may be causing other problems and delays.

Money-Saving Tip:

It is possible to find a number of the plans necessary to build your home online. Research the various plans listed here on different Web sites and print them. In some cases, there will be either no fee or a nominal fee involved. This is worth the investment, though, so be sure to price-shop for affordability and follow the advice of other owner-builders who have used Web sites for this phase. This will save you money in the long run.

6.3: Obtaining Accurate Bids

This is a frustrating part of the process for many because most often, the bids are based on a guess. However, truly accurate bids are not guesses; rather, they are written documents that the contractor stands behind (even if they come in too low). When shopping for accurate bids, the last thing you want is an opinion of what the job might cost. Save the "what-if" phase and guessing for the design phase of the project where you are trying to determine what to include and exclude from the build.

How can you be sure you are receiving an accurate bid and not just a guess? When meeting with the contractor, come prepared with these three key pieces of information:

1. **Timetable:** This outlines each phase of the construction process, the estimated time of completion, and the maximum amount of time the project will take. You will be providing both soft and hard deadlines.

2. **House plans:** This will give the contractor a firm understanding of what his or her job will entail and how much work will be involved. Then, they will know many other contractors and tradesmen will be on the building site and when.

3. **Specification sheet:** Provides the contractor a firm understanding of what supplies and materials will be needed to complete the job. This guideline also makes it clear you do not need anything extra and that you will be price-shopping materials.

With these key pieces of information, the meeting will produce an accurate bid. Be prepared for bids to come in on the high end because contractors have to build in a protection for if you change your mind and when the requirements of the build change. Ask a lot of questions about how the contractor has arrived at this bid, and be as detailed as possible. Request that all numbers and totals be broken down into an itemized list. This will keep you informed, as well as keep the contractor honest.

Having a productive meeting will be beneficial for both you and the contractor, so be sure to have all your information and questions prepared ahead of time. Remember, these individuals are taking time out of their busy schedule to meet with you. It is also a good idea to shop ahead for pricing for all the supplies listed on the specification sheet, not just for the ones you need immediately. This will not only prevent the contractor from bidding too high on materials, but it will also give you an idea of what to budget for. The last thing either of you want to hear is, "I do not know; I will have to get back to you in the next day or two," when question are asked. They will arrive prepared, so you should do the same.

It is important to secure between three and five bids to make an accurate and informed decision; this will take time and can be a hassle. Be patient with the process, though, because it can only benefit you in the long run. With this many bids in front of you, you can make true comparisons among these contractors. Be sure the contractors know every aspect of the job. When the bid comes in, double-check to make sure it includes all the information you provided them. If they left anything out, ask them why. Inquire about any information that has been added, too.

Be keenly aware of the dreadfully low bidder. Often, when collecting a set of bids, one will come in awfully low. You have to ask yourself why before jumping at the chance to save money. Will you be sacrificing quality? Did they include all the information for the entire job in the bid? Do they stand behind the bid, or do they have a reputation for bidding low and adding additional costs at the end? Yes, these are tough questions to ask and to research, but it cannot be overlooked, so get used to performing these tasks. You will get better at it over time.

Doing your homework about each contractor you are collecting bids from is crucial. Ask for references and use them. Find out what they are like to work with, their workmanship quality, and what their billing at the end of the project is like. If possible, visit other sites where they have completed projects. You must be aware of false references, too. It is a terrible business practice, but some individuals still try doing it.

Notes From the Field

If you are too trusting of a person, then you are headed for real trouble. Double-check everything, go through every detail with a fine-toothed comb, and do not be afraid to be suspicious of everyone. You are spending your hard-earned money when you hire these individuals, so you want to be sure your money is being spent well. If all of your research still leaves you feeling uneasy, contact the Better Business Bureau (BBB) to obtain more information and resources, as well as your area builder's association.

Making the final decision about whom to choose can be difficult. Remember, the decisions you make are significant ones, but they can be changed. If something is not working out or an addition needs to be made, it is your right to do so. Be as communicative as possible and remember to add extra cushion in your budget and in the bid for unexpected changes. This will alleviate a lot of stress and headaches through the building process.

Remember, all prices are negotiable. Also be aware, though, that these contractors know what these jobs entail and will bid accordingly. While you might want to try to get contractors to come down on their prices as much as possible, it may not always be wise to do so. You do not want to sacrifice quality by taking the lowest price or by convincing a contractor to lower their price. Some contractors need to stand firm with their bids in order to account for rising fuel and freight costs. That fact must be kept in mind at all times.

Money-Saving Tip:

Though unconventional to some, others believe homeowners should include a penalty clause when writing up contracts for their contractors. These penalty clauses are in place to protect the homeowner from excessive delays and excessive costs associated with being off-schedule. The penalty clause charges a fee to the contractor for each day they go beyond the hard deadline set for the project. The penalty clause could also save you money in the long run if you fear the schedule will not be properly adhered to during the build. Make sure a lawyer draws up or approves any contract.

6.4: Getting the Plans and Permissions

Before making the commitment to draw house plans, it is a good idea to visit your local building authority. Some municipalities have Web sites dedicated to their building authority that can be visited for preliminary information. During this first meeting, ask the person behind the counter:

- Is a zoning map available? This will help you determine if for-sale lots or land can truly be developed. Note where elevations are, where water is marked, and where municipal lines are drawn.

- Can a copy of the local building codes be provided? You will want to provide this to contractors who are not from the area.

- What engineering requirements are needed for the drawings prior to approval?

- Do elevation guidelines need to be accounted for? If so, what are the specifications?

- Are there pamphlets or booklets available for owner-builders to help them through the process? Some building authorities discourage owner-builder construction projects, so be prepared to face some possible snags.

- What is the typical timeline from submittal to approval? Remember to plan plenty of lead-time.

- How many copies of the plans need to be submitted? Find out if you can save time by faxing information in, rather than setting up another appointment for a meeting.

- Do I have to submit a list of the contractors I plan to use? You will have a record of this anyway, so it might be as simple as mailing, e-mailing, or faxing this information to the office.

You cannot build anything anywhere without permission from the local code enforcement office, planning department or, in some cases, a home-owner's association. Patience plays a key role throughout this process, so do not get discouraged. Plan on getting permission at least three months ahead of time; that way, changes and objections will not slow down the build. If your city or town's code enforcement office is notorious for taking a long time, give yourself longer than three months to accomplish these tasks.

Do not attempt to build or install anything without all of your permits in place. Hundreds of people per state are reported yearly to the building inspectors for trying to get away with this. Failure to produce permits upon demand can cost hundreds of thousands of dollars in fines. In addition to the monetary ramifications, there are also safety issues that could potentially not be addressed if the proper permits and inspections do not take place. All of these problems are avoidable if you adhere to the laws. Moreover, you can be required to remove any construction performed without approval.

When having initial discussions with each of the people involved, be prepared for what changes might occur. For example, if you decide to alter the size of your deck or omit a window from the foundation, what permissions need to be changed? Anticipate that the people you need to meet with have busy schedules as well. They will get back to you, but it may not always be as soon as you would like. Discouragement may be your first feeling, but you have to be accepting of these schedule issues.

So, are all your ducks in a row? Before applying for a permit, it is essential to meet these four jurisdictional requirements (in all 50 states):

1. Ten-digit assessor tax parcel number
2. Legal description, in full
3. A scaled drawing of the site plan
4. Estimate of the costs associated with construction and the entire building process

Notes From the Field

This is normally the point where owner-builders start feeling the pings of impatience. The paperwork process is a lengthy one, but all of it is necessary and cannot be glossed over under any circumstance. If this paperwork is not all in order, you will be delaying your building project. Contrary to popular belief, these details cannot be overlooked under any circumstance. Do not let anyone try to convince you otherwise.

At each of these meetings, you will need to come prepared with all the plans and spec sheets as you did when obtaining bids. The code enforcement will be determined once the plans are presented, which is often the lengthy part of the process. The people you are seeking approval from will need to be certain that the builder can easily follow these plans. They will ask a lot of questions that you must be prepared to answer at the time of the meeting. Otherwise, you will be faced with waiting for another meeting to be scheduled that will further delay your timeline.

Be sure to prepare accordingly for the fees associated with obtaining your permits. These fees can be high, but your local code enforcement officer will break down how they are assessed and what they are for. Some of the fees addressed during the application process are land use, school impact, road impact, and parks. Some offices allow you to come in after the meeting to drop off a payment and pick up paperwork without the need to schedule another appointment. Find out ahead of time if you will be able to explore this option.

Money-Saving Tip:

Set aside money in an escrow account to prepare for permit fees. Because these fees can be excessive and permits cannot be issued until paid in full, it is crucial to be as prepared as possible. The permit process is long enough without having to worry about coming up with the money for the fees associated with them.

6.5: Finding the Right House Plan

How do you know what the right plan is for you? First, let us discuss where to find the house plans in the first place. There are many plans available at the bookstore, the local library, and on countless Internet sites. You may also choose to talk to an architect or a draftsman to obtain a set of custom plans.

Starting out on the Internet is your best bet to gain a feel for what your options are. There are thousands of styles available, so this can easily become overwhelming. Let the Internet searches you conduct act as a guide for the different features and ideas you may not have considered. It is a good idea to enter this situation with a basic idea of your house visualized in your mind or sketched out on paper. A good place to start is **www.houseplans.com**.

The next step is to compare the house plans you have found with the list of requirements you have set up. These requirements must fit the house plans, so do not skip this step. Remember, liking the exterior of a particular house does not mean you are necessarily going to like the interior it comes with. Consider every option during this process so you do not disappoint yourself later. The last thing you want to experience is regret when your new home is complete. Ask yourself the following questions when sifting through house plans:

1. Do these plans fit my requirements?
2. Can these plans be modified if they do not fit my requirements?
3. Will the changes I want to make cause problems with the construction or structure of the home?
4. Do I need a draftsman or an architect to help me with significant changes?
5. Are engineering drawings of the floor system available with the house plans? If not, can they be obtained elsewhere?

Answering these questions will help you determine if the house plan is right for you. Once you have found what appears to be the right design, go ahead and make a purchase. Checking and rechecking every detail of the house plan is crucial during this selection process, so be as meticulous as possible. If you feel you are going to make any mistakes at all during this process, it might be best to hire an architect and consult with him or her often.

> ### Notes From the Field
>
> Each time you see a house style you like, take a picture of it right away. Keep a camera with you at all times so you do not miss out on these photographic opportunities. Each time you are perusing a magazine or a catalog and see a picture of a house style you like, cut it out immediately. Keep all of these materials together in a file or scrapbook to help narrow things down during the planning process.

Do not let the process overwhelm you, and be sure to take your time. Choosing the right house plan for you and your family will be time-consuming, so prepare for that. In addition to considering the questions about building plans, note how the house will be situated on the lot, what the natural landscape of the property is, and how future buyers will respond. The needs and lifestyles of the family members living in the home also need to be taken into consideration, so ask the following questions:

1. If you are a new couple, do you intend to have children?

2. Do you need a home office?

3. Do you need a guest room?

4. Does this plan allow for future expansion?

5. Should there be a separate dining room or other entertainment area?

6. If you are currently parents, is there enough room for the family to grow up in the house, rather than grow out of the home?

7. Is there enough room for holiday gatherings or other social events?

8. Can future rooms or bedrooms be built in the basement and meet code?

9. Do you need a separate laundry room?

10. Is there a loft space available in the attic or above the garage for storage?

Money-Saving Tip:

Shop around. If you are unhappy with the pricing of a house plan, but like the plan itself, shop around to see how to cut costs on drawing them up. There is a wealth of resources online about affordable house plans from home builders, contractors, and architects. You can also find books with specific house plans at most bookstores and on bookseller Web sites.

CHAPTER 7
Heating and Cooling

7.1: Types of Heating

There are many heating choices, each with its own benefits. It is imperative to make an informed and educated decision based on your needs, your budget, your preferences, and the information you have gathered. Do not base your decisions on commercials or advertising hype, a well-intentioned family member, or a friend who works in the industry. Base your decision solely on the information you have gathered and discussed with others you will be sharing the home with.

Wood-burning stoves and fireplace inserts are much more efficient than fireplaces because of heat loss. A pellet stove is efficient, warm, and thermostatically controlled. However, there are two negative features: If the electricity goes out, so does your stove because pellet stoves use a motor to feed the pellets; the second feature is it still has to be filled every six to eight hours and use a minimal amount of electricity. The more costly pellet stoves combine the convenience of a pellet stove and ability to load logs when the electricity is off, but there is no thermostatic control with logs. This principle applies to log and coal stoves as well, with no regulation of heat and more frequent loading of fuel than pellets. Coal can also throw off toxic fumes and is generally not recommended for indoor use, just as gas heaters are not either.

Here are some choices of energy sources to explore:

Electricity: Electricity is readily available most anywhere, but comparison-shop how much it would cost to run electric heat through a brutal winter in contrast to the other options. In warmer climates, this will be less of an issue; you will look at the cost of cooling instead. Plus, if power should get knocked out during a storm or for any other reason, you will be without heat if you do not have a back-up plan in place. However, this is a clean fuel alternative that requires little maintenance.

Fuel oil: Oil is a heating option available everywhere. With regular maintenance, this is another relatively clean fuel choice. There are more disadvantages to using fuel oil in comparison to electricity, given the need to store the oil tank on your property, either in the basement or outside next to the house. Also, you will need a chimney. However, oil is not a renewable resource, which makes it an unfavorable source for green builders.

Gas: Because gas comes in two forms, liquid and natural, its availability is based on which you choose. Natural gas is dependant on how close you are to gas pipelines, and that in turn is based on geography. There is no storage required. Liquid gas is also a clean fuel option but mandates on-site storage since it is delivered. Most building codes now require in-ground tanks outside. Like oil, liquid gas may require a chimney to be installed. Although it is one of the cleanest fuels available, no matter which form gas is in, it is not a renewable resource.

Coal: Though once a popular heating choice, coal is rarely used anymore and, therefore, is less readily available. An exceptionally dirty fuel before, during, and after burning, coal has an above-average risk of house fires associated with its use. Like oil and gas, it is not a renewable resource. This heating option is rarely recommended when speaking with architects and contracting specialists.

Wood: The availability of wood depends on your location and the supply of hardwoods (oak, ash, maple, cherry, and elm) in your area. The advantages of heating your home with wood are numerous, including the ability to maintain and grow your supply on your property. This fuel is a renewable resource, but substantial labor is required. This use of wood, of course, requires the installation and maintenance of a chimney. In order to avoid the risk of fires, chimneys must be cleaned professionally every year or two depending on usage. Using natural, untreated wood cuts the menace of chemicals adhering to the smokestack lining. Green builders favor this heating option so long as clear-cutting is not involved and replanting takes place.

Solar: This heating option is solely dependent on how much sunlight is available. This is a clean fuel option, and there is no charge. Due to the unreliability of weather and the inability to store large amounts of energy without substantial costs, a back-up system is required for those choosing solar heat. By far, this is the most favorable choice for green builder.

Money-Saving Tip:

Choose the most cost-effective type of heating for your build. In terms of fuel, natural gas is less expensive than heating with oil or electricity. If you have enough property to grow your own fuel supply that would be your best choice to make. Do your homework.

7.2: Types of Heating Systems

Forced-air heating units

There are a lot of advantages to using a forced-air heating unit, including no worries of frozen pipes on freezing winter days. Forced hot-air units use air as the heat-transfer medium. Ducts are used to transfer the heated air throughout the home through a ventilation system. Forced hot-air systems are known to be the most commonly installed heating unit in the United States. There are several advantages to using forced hot-air heating units:

- The ductwork is compatible with central air conditioning units.
- There is no risk of water leaking and causing damage.
- They are energy-efficient in comparison to other heating units.

There are some drawbacks to using a forced-air heating unit that you should be aware of as well:

- If the forced hot-air unit is not installed properly, air leaks can occur in the duct work.
- These heating systems tend to be noisy when their fans are running.
- Filters must be used and replaced periodically in order for the heating unit to run efficiently.
- There is a chance of distributing cooking odors and allergens throughout the heated space when the heating unit is in operation.

Hot water heating units

These heating systems are either gravity or forced hydronic. Because gravity systems are found mostly in older homes and are normally replaced due to their inefficiency, the most popular choice is forced hydronic. Forced hydronic systems are commonly heated by a boiler system that is run by either gas or oil.

Heat pump units

These heating systems draw air from outdoors and circulate it through the home through ducts. During the warmer months, it does the exact opposite by drawing the hot air out of the home and bringing it outdoors. There are many benefits to using a heat pump unit, including:

- Because heat pump units do not use fuel, they are safe and clean to use.
- Heat pump units are able to deliver a maintained airflow, rather than blasts of hot air like a traditional heat unit.
- Heat pump units do not produce dry air, so a healthier environment is maintained.
- Heat pumps are more energy-efficient and cost-effective.
- Because heat pump units run throughout the year, they tend to cost less than running a separate heating and cooling system.

Solar units

At one time, solar energy was thought to be out of reach and practicality for the average homeowner. However, modern advances have made the solar collecting panels affordable and durable. There are several possibilities, depending on where you are geographically. The Southwest, where the sun shines most of the time, has in the last decade added passive solar to their arsenal. Passive solar heating requires windows facing the south and brick floors to absorb the sunlight. Other types of passive solar heating are similar, with variations of the sunlight striking collectors at the base of windows and storing the heated water for the evening when a pump is turned on and the water circulates under the brick flooring to supply radiant heat. Other methods involve solar panels on the roof or behind the building, also facing south. If conditions are met, this can save significantly on hot water, even if it is not used for heating the home.

Electrical thermal storage units (ETS)

These heating systems offer cost-effective heating options for homeowners by using electricity during off-peak times. High-density ceramic bricks store the heat during off-peak hours, allowing for heat transfer to occur 24 hours per day. Maintaining the desired temperature throughout the day is achieved through use of a thermostat. How many heaters are needed depends on how many rooms are in the home, how well the home is insulated, how many windows there are, and the square footage of the home.

Radiant floor heating systems

These heating systems are available in both hydronic and electric forms. They heat the room from the floor up, so it is an energy-efficient choice. Special tubing is installed into concrete, and heated fluid runs through this tubing to heat the room. Heat sources for this liquid can come from solar energy, heat pumps, wood stoves, demand water heaters, and regular water heaters. Radiant heat is also available in electric form, with wires and conduits under the ceramic tile. Electric radiant heat is efficient but can be expensive depending on your location. There have been recent efforts to draw electric solar power to supplement "on-the-grid" from the electric company or generator-powered electricity.

> ### Notes From the Field
>
> Be sure to hire a reputable installation contractor with significant experience. You want the heating unit to run efficiently and be easy to maintain. In addition, no one wants the lifespan of the unit to be compromised as a result of poor installation.

Every home heats differently, so choosing which is best can be difficult and confusing. Your best bet is to speak to manufacturers, installers, contractors, and tradesmen individually to obtain the best information. The more educated you are about these heating options, the better chance you will have at making the best decision. Talking to other homeowners about their heating choice will assist during the decision-making process.

Money-Saving Tip:
Thermostat settings can reduce energy bills by 10 percent yearly. Simply set the thermostat for 60 degrees while you are sleeping or away, and 68 degrees when you are home if you live in a cold climate.

7.3: Types of Cooling Systems

Cooling systems in homes with forced-air heating will likely be sharing the same ductwork as the heating system, so everything is readily adaptable for use. Because the heating and the cooling run off the same pump, no additional equipment is needed. This makes for good, cost-effective installations. Conversely, if the heating system installed does not include ductwork, a separate system will be required, as with baseboard heating.

When the time comes to make a decision about the air conditioning unit, it is important to take note of these crucial points:

1. To reduce the noise of the compressor, avoid installation of cooling units near decks, patios, bedroom windows, and dryer vents.

2. Pay attention to the Seasonal Energy Efficiency Ratings (SEERs), which are the basis of comparison for each cooling unit.

3. For those deciding not to install air conditioning right away, be sure to keep the furnace area roomy enough for a cooling unit to be installed easily and comfortably.

4. For those deciding not to install a heating unit requiring ductwork and who also want to wait on installing a cooling unit, be sure to install the ductwork for air conditioning if the decision is changed in the future.

5. Avoid choosing a cooling system that is too large. This will cool the air quicker, but it will leave the air quality poor and clammy. In addition, it will turn on and off more frequently because of the rapid changes.

6. If plans of expansion are in the future, be sure to decide upon a cooling unit large enough to accommodate those changes.

7. Be sure the cooling unit is energy-efficient, has replaceable or cleanable filters, and is designed to keep dirt, leaves, and debris from clogging anything.

8. Be sure there is proper drainage available in the cooling unit for condensation build-up.

9. Be sure to select a cooling unit that does not have fixtures, parts, or refrigerant that is being discontinued in the future due to environmental regulations.

About Central Air Conditioning Systems

A central air conditioning system is either a split-unit or a packaged unit. In a split-unit, there are two cabinets. The first cabinet contains the evaporator and is housed indoors. The second cabinet, which is constructed of metal, is located outdoors and contains the condenser and the compressor. Some split-units also have indoor cabinets containing a furnace or heat pump. For packaged units, all the components are in a cabinet that is mounted either on the roof or on a slab outside of the house.

Notes From the Field

If you choose a central air conditioning unit, be sure all the interior ductwork is well-insulated. Just as it is important to keep the hot air blowing through it for heating, the same concept holds true for cold air. Poorly insulated ductwork will cause the air temperature to rise or fall depending on the external temperature.

Central air conditioners are more efficient than individual room air conditioning units. In addition, they are quieter, out of the way, and require no storage. Keep in mind that the unit will not operate efficiently if it is not sized properly for the home. When hiring a contractor, the following specifications should be required:

- Allowances for space indoor and an access door or panel to clean the evaporator coil are made.

- Duct-sizing methodology must be used to match the volume of air to the duct size for more efficient delivery.

- There are enough delivery, as well as return, registers.

- Ductwork in the attic should be avoided to prevent temperature changes.

- Make sure the contractor heavily seals and insulates ductwork.

- The condensing unit is in an area where the noise is not a nuisance and is accessible for repair and maintenance.

- The refrigerant used matches the manufacturer's guidelines.

- The thermostat is located away from all heat sources.

There are four kinds of individual-room air conditioning units available on the market today for those who choose not to use central air conditioning:

- **Window mounted:** This is the most common type of air conditioning unit. Like with all things, there are both pros and cons to this choice. Window-mounted air conditioners are easy to install, and they work with the wiring in most homes. But they are designed mainly for double-hung windows, and if you live in the North, you will need to remove them at the end of the season.

- **Wall mounted:** These units are permanently mounted into the wall. Some are cooling only, while others offer both heating and cooling capabilities. Wall-mounted air conditioners are easy to repair, easy to replace, and no storage is necessary. But they can be costly to install, and they require a metal sleeve. In addition, they cannot be removed at the end of the season if you live in a colder climate. Over time, as the house structure settles, there could be the risk of air leaks.

- **Window/wall mounted:** These units are versatile because they can either be mounted in a window or permanently installed in the wall. These units are portable, and you have the ability to move it if you are unsure you want a permanent unit. But there are no window installation kits included; plus, you may need a dedicated 220-volt outlet.

- **Portable:** This unit comes on wheels and can be brought from room to room. These units are efficient, are about the size of an old-fashioned radiator, are easy to store at the end of the season, and are very good dehumidifiers. But, they eat up floor space and must be vented out a window. These units will also stop operating if the fluid reservoir of the humidifier fills up and needs to be drained to a bucket or other source. Finally, portable units are more expensive than the conventional types.

- **Evaporative cooling:** These systems are designed for those planning to build in hot, dry climates. Evaporative cooling systems are available in window units, as well as central units, and cost considerably less than traditional cold air systems. They were once a popular source because of high-electrical efficiency; they are still used, but to a lesser extent. Often referred to as "swamp coolers," evaporative cooling systems have a fan that blows a fine mist of water into the air, providing enough moisture to cool the skin and air. It can drop the heat about 15 degrees. But, when the temperature exceeds 100 degrees (Fahrenheit), they are rather ineffective. Additionally, due to the increase in urbanization, fewer areas, even in the desert, are totally free from humidity. Last are the environmental concerns with the large amount and continuous need for water. The units themselves are high-maintenance because there are screens that are kept wet for the fan to blow through. More often, this kind of air cooling system is in the desert where sand accumulates and cakes on the filters. They need to be cleaned and serviced throughout the season.

Money-Saving Tip:

Close off the fresh air vent during cooling; that way, the outside air is not being cooled. If the air outside is cooler, then open the vent to allow the fresh air to come in.

7.4: Air Cleaners and Humidifiers

Electronic air cleaners are best used with forced-air heating and cooling units. Not only do they clean the air, but they also tend to make the air smell fresher. There are numerous benefits to installing air cleaners:

1. They remove up to 95 percent of dust and bacteria and in some cases, also viruses.

2. They allow for a cleaner home.

3. They reduce allergens, ultimately making the home more comfortable for everyone.

The Environmental Protection Agency (**www.epa.gov**) has put together a checklist of points for consumers to take into consideration before making the decision to add an air filter to their home:

- **Installation:** In-duct air cleaning devices have certain installation requirements that must be met, such as sufficient access for inspection during use, repairs, or maintenance.

- **Major Costs:** These include the initial purchase, maintenance (such as cleaning or replacing filters and parts), and operation (such as electricity).

- **Odors:** Air cleaning devices designed for particle removal are incapable of controlling gases and some odors. The odor and many of the carcinogenic gas-phase pollutants from tobacco smoke will still remain.

- **Soiling of Walls and Other Surfaces:** Ion generators generally are not designed to remove the charged particles they generate into the air. These charged particles may deposit on room surfaces, soiling walls and other surfaces.

- **Noise:** Noise may be a problem with portable air cleaners containing a fan. Portable air cleaners without a fan are normally much less effective than units with a fan.

Humidifiers, just like air cleaners and central air conditioning, can be installed directly into the ductwork of the heating unit. To do so, they must be hooked into an existing water supply to deliver necessary moisture into the air. There are several benefits to installing a humidifier, some of which include:

1. A comfortable level of moisture is maintained in the air, which is good for respiratory health.

2. They make harsh, dry air easier to breath.

3. Installation is simple, particularly if existing ductwork is in place.

4. Condensation buildup on windows is reduced or eliminated; therefore, mold and mildew will not breed.

Notes From the Field

Be aware of issues with freestanding humidifiers spraying warm mist. Some homeowners have reported the growth of black mold after operating one of these units. Keep this reality in the front of your mind when deciding if a freestanding unit or an installed unit would be the best choice.

Forced-air humidifiers come in three different styles: disc wheel style, flow-through style, and drum style. The drum, while the most popular, is not the most efficient. It is composed of a sponge or wick that absorbs the water, and the air blows across. The problem is that the material of the pad becomes encrusted with sediment and must be replaced according to the hardness of your water supply. The flow-through has a constant stream of water flowing across a screen, connecting to a drain pipe for the excess. It has fallen into disrepute given the extraordinary waste of water. The rotary disk uses the basic premise of a drum but, instead of pads, it utilizes about 40 disks that spin and atomize the water. The maintenance is considerably less than for a drum type humidifier.

Money-Saving Tip:
Using a humidifier makes you feel warmer in colder months. This will allow for a lower thermostat setting, thus saving on energy costs. Roughly 20 to 40 percent humidity is recommended.

7.5: Ceiling Fans

Use of a ceiling fan offers both heating and cooling advantages. It has been proved that in the summer months, people are 4 degrees cooler when a ceiling fan is on. In the winter months, ceiling fans successfully distribute heat from room to room. This is particularly true for small rooms containing space heaters and wood stoves. People often choose to use ceiling fans on days that are warm, but not hot enough to run the air conditioning.

By using a fan when expedient, homeowners find they are being cooled as well as reducing energy costs. Most ceiling fans use about as much electricity as a 100-watt light bulb. When used properly, a ceiling fan can save energy bills both during winter and summer months. In the winter, the fan should turn in a clockwise direction to bring the heated air away from the ceiling; this causes about a 10 percent cost savings. Using fans in a clockwise direction during the colder months forces the rising hot air from your heating system back down to warm you, resulting in similar energy savings.

Using fans during the summer months brings the biggest savings. Reverse the direction the fan moves so it turns in a counter-clockwise direction and it will cool the room between 4 and 8 degrees. This cooling reduces air conditioning costs by up to 40 percent.

To reap the most benefits from ceiling fans in your new home, use them in several rooms throughout the home. Rooms that are the most frequently used, like kitchens and family rooms, should contain ceiling fans; this will ensure comfortable air temperature throughout your home. There is a caveat, though: When it is warm, turn on the fans only in occupied rooms. Ceiling fans cool by increasing the evaporation on your skin and it is a waste of energy to run fans when there is no one to cool. Fans should be used to force the warm air down from the ceiling toward the layers of cool air beneath. You should be able to reap significant benefits by lowering the thermostats and achieve greater comfort by circulating the warm air through all used rooms.

When ceiling fans are installed in bedrooms, it creates an environment for a more restful sleep for some. The white noise from the movement of the fan has also been said to be relaxing. Using a ceiling fan year-round at various speeds can help those who cannot sleep in a still and stuffy room.

While some may feel it is unconventional to have a ceiling fan in the kitchen, others have found it beneficial, particularly during holiday cooking. Not only does the ceiling fan disburse the hot air generated from the stove and oven, it also distributes hot air throughout the rest of the house during winter months.

Ceiling fans normally come in two different sizes: 42-inch fans for small rooms, and 52-inch fans for larger rooms. You can also find fans between 30 and 36 inches for very small rooms and also 56 to 60 inches for extremely large rooms. The most commonly purchased fan size is 52 inches. If you are unsure what size ceiling fan you should choose, speak to a specialist at your local home improvement store.

Notes From the Field

Be sure to choose a fan that fits in with the décor of the rest of the house the first time around. Otherwise, you will be facing unnecessary expenses making another purchase and hiring an electrician to perform another installation.

When installing the fans, be sure they are between 7 and 9 feet from the floor and 10 to 12 inches from the ceiling for optimal use. Fans should be 2 feet or more from any wall, door, or cabinetry. This can best be achieved by placing the fan directly in the center of the room. If the fan is being installed in a narrow entryway or hallway, this might not be possible.

Money-Saving Tip:

Use a fan in conjunction with an air conditioning unit. The fan will circulate the cool air throughout the room. You can install the fan, and they do not necessarily require the talents of an electrician.

7.6: Fireplaces

Fireplaces, which were once used for lighting, heating, and cooking, have become less functional. While they are still primarily used for heating, there are many decorative options available for fireplaces. Although they are no longer considered a necessity due primarily to modern utilities, they are still desired and appreciated by many homeowners.

Today, fireplaces fall into two categories: those that either heat a room or an entire home, and those installed merely for ornamental purposes. Homeowners tend to choose the ornamental fireplaces over the functional ones, despite their reputation for not being efficient in terms of heating ability and related costs.

There are several disadvantages to consider before installing a fireplace. These include:

1. Fireplaces experience a loss of heat through dissipation up the chimney.

2. Warm air still escapes up the chimney and outside, even when a fire is not lit and the fireplace's damper is closed.

3. Brick and stone do not insulate well, so there is a chance of having cold spots in the wall where the fireplace is located.

4. Unless serious design considerations are made, a fireplace takes up valuable wall and floor space.

5. Professional chimney cleaning and regular maintenance is required when a fireplace is in regular operation.

6. Fireplaces can cause safety concerns, particularly when small children are present.

7. Fireplaces can be messy when the fires are burning and with the soot left behind when the fire is put out.

Masonry fireplaces are the original type and are still popular today. These types of fireplaces must be built from scratch on the building site, which translates to high contractor costs. Masonry fireplaces increase the salability of the home because they are attractive, carry a good reputation for being well-built, and there is little maintenance required once they are installed. Original masonry fireplaces contain firebricks, which can handle excessive heat when wood is burning against them.

Another type of fireplace is a masonry fireplace with a steel firebox installed. The only difference between these fireplaces and the original masonry fireplaces is prefabricated steel boxes are installed in place of the firebricks. They are less costly than the original masonry fireplaces, but retain the same attractive appearance, low maintenance, and heating capabilities.

To reduce construction costs for those who are interested in installing a fireplace, a prefabricated, built-in fireplace is recommended. These designs can be installed virtually anywhere in the existing home. Because these fireplaces are lightweight, there is no need to install any expensive masonry units.

Notes From the Field

Some owner-builders have reported how the installation of prefabricated fireplaces has suited their fickle moods well. If they changed their mind about having the fireplace in the room or where it was located in the room, a costly renovation was not necessary. Others were able to add this in if there was no room or the installation during the initial build, as well, without compromising the room's construction or running into high costs.

Another cost-effective fireplace, and the least expensive of all the options, is a freestanding one. These fireplaces are available in designs to suit just about any décor ranging from country to modern. Colors and the materials they are constructed from also vary. For example, these fireplaces are characteristically constructed from soapstone, steel, cast iron, brass, and other types of stone. In terms of placement, these fireplaces are not built into or attached to the wall. However, they do require chimney access.

Wood-burning fireplaces, otherwise known as woodstoves, are an option for those who are looking for function over design. These fireplaces tend not to be ornamental, but provide the most effective heating of all the options. Woodstoves are available in a wide range of styles and sizes, so they are able to fit any décor. Because woodstoves create a lot of heat, locate it in an area of the home where heat is needed the most, like the center of the home, so heat is distributed evenly throughout the house. Materials used to construct woodstoves range from steel to cast iron to soapstone.

Fireplace efficiency depends completely on design, construction, and installation. What is burned in the fireplace is also significant; for example, never burn any garbage, chemically treated materials, or plastics. These items can be explosive and cause build-up that could eventually catch on fire. It is advised to burn only mixtures of maple, oak, pine, and fir.

Fireplace locations should be determined in the original set of house plans, or in the case of fabricated models, once the house has been completed. It is beneficial to consider all elements of the home's design scale when deciding where to position the fireplace because once it is installed, it cannot be moved. It is advised that fireplaces be located on an interior wall of the home. This will prevent up to 20 percent of heat loss through the masonry mass.

When taking the chimney into consideration, there is much to prepare for. You have to plan for the chimney walls, height, lining, mortar, damper, cap, flashing, sealing, and cleaning. Also, bear in mind that no matter where it is positioned, the chimney is a prominent feature, both inside and outside the house. For the best heating efficiency, it is best to house the chimney in an exterior wall, rather than slinging on the exterior wall.

Hearths are created as a means to protect the floor of the room where the fireplace is located. Covers and screens prevent sparks and exploding embers from causing a fire in the room where the fireplace is located. Mantles are optional, ornamental features that are often found on masonry or built-in fireplaces and are a traditional feature that can be constructed from most any sturdy material. Some fireplaces have ash pits and some do not; it is all a matter of preference and whether there is enough depth at the base of the fireplace for the ash pit.

Direct-vent gas fireplaces are designed for those who want to have a fireplace without having to cut wood, store wood, or perform the necessary cleaning involved with traditional wood-burning fireplaces. Direct-vent gas fireplaces have many advantages over traditional gas fireplaces in terms of efficiency, versatility, and safety. Combustion air is drawn from the outdoors, rather than using conditioned indoor air. Some direct-vent gas fireplaces require no flue, and their pipes can be installed directly into the wall.

Fireplace energy efficiency can be increased several different ways:

1. **Install a fire back:** Fire backs are a sheet of metal that sits in the bottom of the hearth and leans against the back fireplace wall. They are available in both stainless steel and cast iron, and are efficient because the heated metal radiates into the room.

2. **Glass doors:** Installation of tightly fitted glass doors will reduce the loss of inside air, but it will not be 100-percent effective. The reduction of air loss through this installation makes it worth the expense.

3. **Tubular grates:** Using these in conjunction with a blower will increase the fireplace's efficiency, delivering the same type of radiant heat as the fireback.

4. **Heatilator:** This is the most expensive option, but it is also the most efficient. The shell within the construction fits into the fireplace and distributes heat around the firebox. This heat is then delivered into the room through use of a vent, with or without the help of a fan.

5. **Outdoor intake:** Installation of a small pipe used to pull air from outside, rather than the warmed inside air, will increase fireplace efficiency.

Money-Saving Tip:

If your home's design calls for frequent use of the fireplace, consider adding fireplace inserts, doors, or covers to the design. This will reduce heat loss when the fireplace is used in the home, thus saving on heating and energy bills.

CHAPTER 8

The Beginning of Your Castle (Phase One)

8.1: Permits, Building Codes, and Inspections

Most cities and counties will not allow for any kind of construction on a piece of property without a building permit. The type, details, and location of the construction all need to be approved to obtain this permit. The required sanctions are obtained through government offices; without such, nothing can be done to initiate a build. It is critical to start this process in advance so it does not interrupt your proposed groundbreaking and start date.

You may, however, be able to begin clearing the trees and brush from the building site so long as heavy equipment is not brought in. Because building codes and laws vary so much, check on this detail before firing up the chain saw and gassing up your four-wheeler. Avoiding fines and delays should always be in the front of your mind. Do not be smug and assume no one will know; there could always be a neighbor or nearby worker willing to notify authorities.

The different permits can be confusing. Visit the code enforcement officer or equivalent person in your city or county. This individual, or others working in the office, can clarify requirements so you understand them. They will explain what permits are needed, the fees associated with each, and where the document needs to be located on the property during the build. Workers in these offices are usually fountains of knowledge, so ask a lot of questions, and do not feel stupid; they have heard them all.

Notes From the Field

You may find the clerks behind the desks at town halls in small towns to be armed with more information than those in the busier metropolitan areas. This is normally because they typically have more time on their hands to have discussions with residents, the code enforcement agent, and appraisers.

The application normally consists of one or two pages. They must be accompanied by a complete set of drawings and the surveyor's plans, showing the locations plus orientation of the house. Fees associated with the permit are characteristically based on the square footage of the home. These fees can be itemized into an understandable list upon request if it is not already available to you.

Once the building permit is issued and construction is complete, officials perform an inspection to determine whether all the codes were followed properly. If you have any questions about this process, ask. Contact your state's regulatory agency or your local building department. It is not possible to have too much information, but it is possible to not have enough. Do your research well.

Money-Saving Tip:

When you visit the code enforcement office, request a copy of all the codes that may pertain to your building project. If they cannot produce a copy, ask them where you can search on the Internet for this information. If there is no information online, request if they can speak to you personally about the codes so you can write everything down. Having this information at your fingertips or in your files will save you money.

8.2: Groundwork and Excavation

You are now ready to begin the construction process. Before starting anything else, a driveway must be created to gain access to the building site. Check with state and county transportation authorities to see if there are restrictions or guidelines about where the driveway must be on the property. If you do not check first, you might waste time and money tearing it apart and doing it all over again, in addition to the possibility of getting a fine.

Clearing the lot is the next step of the groundwork and excavation process. This, of course, involves the removal of trees, stumps, and brush that can cause interference with the construction of the house and the driveway clearing. Removing as much of this as you can on your own will save money in subcontracting fees.

Notes From the Field

Plan ahead! If you decide to utilize a wood stove or fireplace, cut the trees you are going to remove into stovelength pieces and stack it out of the way on the edge of your property. With heating fuel prices on the rise, you will not regret attending to this detail. You can speed up the process by renting a wood splitter. If you do not feel comfortable storing the wood on your new property due to theft or other issues, haul it to your current residency or to a family member's home until you are ready to move on to your property.

Be sure to check if environmental laws are in place regarding sediments and other runoffs into streams that might occur from the construction site. It is crucial to follow these codes to protect the streams and other environmental factors. If you ignore this detail, you will find yourself facing fines and delays, public disdain, and perhaps upset neighbors.

Before the excavation can begin on the property, be sure to meet with the excavator ahead of time to mark off where the house is going to be situated. This is where the foundation, slab, or crawl space will be dug. The only thing that needs to be done during this meeting is physically marking the corners of the foundation; the excavator will do the rest.

Setting up a timetable for this phase of the project depends on the type of foundation being dug, the slope of the land, and any other preparation that needs to occur on the building site. Be sure to pay close attention to what the excavator is doing so the hole is being dug properly to ensure the house is going to be situated as in the drawings.

Money-Saving Tip:

When hiring an excavator, hire someone local, if possible, as they are most familiar with the terrain. Speak about the additional costs associated with the job such as removing large rocks or blasting through ledges. Knowing this information in advance will help with budget preparation, save time, and prevent cost-associated delays.

8.3: Foundations, Footers, and Slabs

Despite the expenses basements pose in terms of installation cost, homeowners choose to include them in their plans to allow for more space. Basements are often converted into family rooms, recreation areas, and other utilitarian uses within the home. In some cases, given the slope of the land, it is possible to add a walkout from the basement to the yard. These additions and renovations make the home feel larger. Basements are often considered another level or floor of the home.

A footer, as the name implies, is the concrete base supporting the structure. If this item of the construction is built to be solid, sturdy, and stable, the rest of the house will sit firmly and be positioned well. Be conscious of where the frost line is located because the footing should extend below this point. A frost line is the depth at which the water in the ground freezes, and it depends on the soil content as well as elevation above sea level. If not properly located, the footing and foundation will crack as the ground shifts. Your foundation contractor should be able to explain this, along with answering any questions you might have.

A house needs a foundation or slab to support its especially heavy structure. The stability and reliability depends on good site preparation. A foundation or slab also places a barrier between wood materials and the ground. If the raw lumber came into direct contact with the ground, rot and termite infestation could occur.

Notes From the Field

If the water table on your building site is particularly high, you will need to make sure the foundation contractor accounts for this. Otherwise, you will be faced with water leaks and a wet basement. Some contractors claim to be knowledgeable when it comes to dealing with tricky land grades but actually are not, according to reports from some owner-builders. This is a good example of why it is of the utmost importance to check references at all times.

Concrete is the most common material used for slabs and foundations. The fundamental layer of homes has been known to be made of stone, brick, treated lumber, and concrete blocks, though most often these materials are more common in older homes. There are three different kinds of concrete foundations: poured foundation, concrete blocks, or post and pier. Concrete blocks are just as strong as poured concrete when they are filled with gravel and cement; however, this material is less often chosen because it is more labor-intensive. Post and pier construction lacks a foundation pad or perimeter footing. It is typically used in tropical regions and is used for the cheapest type of buildings; it has been outlawed in most jurisdictions within this country for safety reasons. A poured foundation tends to be the most popular choice among homeowners and owner-builders.

When working with a foundation, pouring a slab is also known as pouring the concrete floor for the basement. Steel rebar is installed within the concrete to make the structure stronger. Some homeowners request adding extra rebar beyond what is recommended to ensure long-lasting stability and strength. Do not be afraid to request information about this option during the bidding process because foundation contractors are familiar with how common this practice is.

More often than not, a service professional installs the foundation for new construction. He or she needs to come in and assess the land for proper preparation before any work can begin. This assessment includes learning the slope of the property. This type of planning allows the professional to design a foundation so water will drain away from the concrete and prevent moisture build-up.

Proper drainage is crucial, particularly for those who are building on land containing a lot of water. Be present at the time of the assessment to answer any questions the foundation contractor has; this will also be a good opportunity to ask any questions you have. This way, you can gain an early understanding of their work ethic.

This service professional will also look at what kind of soil is present at the building site. This will determine what kind of drainage system needs to be installed, as well as how much pipe drainage needs to be installed. For example, if the lot is comprised mostly of sand, it will require a different drainage system than one that contains considerable amounts of clay. Water moves slower through clay than it does through sand, so it needs to be treated differently. If you notice that little or no attention is paid to these issues, ask why. As with other contractors, get more than one bid and compare attitudes and attention to detail. You may get different answers.

Some owner-builders choose to have insulated foundations installed in their new construction on the exterior walls. The benefits of having an insulated basement, slab on grade, or crawl space include savings on energy bills and below-grade rooms made more comfortable. There are also moisture-control benefits where condensation would normally be an issue. More benefits include:

- Reduction of heat loss through the foundation walls
- Damp-proofing coating is protected during back filling
- A barrier against moisture
- Protection of the foundation during freezing and thawing in particularly cold climates
- Reduction or elimination of condensation on the surfaces of the basement

A less costly option is to insulate the interior walls of the basement. Some of the benefits to choosing this option include:

- It can be installed into existing buildings and is less expensive to do so
- There are more materials to choose from
- There is no threat of insect infestation, unlike insulating exterior

foundation walls

• The space is isolated from the colder ground more effectively.

The best way to approach this phase of the construction process is to ask yourself what the best foundation is for you. The type of house you are building, the location of your property, and your budget are a few of the factors that will play into this decision.

8.4: Underground Plumbing

Remember, once all the walls are up and the foundation is finished, it is difficult to go back and add plumbing elements later; be sure to take your time during the planning phase; otherwise, you will experience a lot of headaches later. If, for example, you want a pool a few years down the line, accommodate for that plumbing now.

Once a list of all the necessities and extras is made, you will go through the bidding process for a contractor similarly to the other contractors you have needed along the way. When the selection portion of the process is complete, it is best to meet with your contractor face-to-face to discuss the project in detail and firm up completion dates as best as possible.

Notes From the Field

Keep in mind; like with the other contractors you have been through the bidding process with, you must be aware of the low bidder and the false references. If you do not remain aware of these realities, you could be headed for big trouble during the course of the build. In some cases, license information is all you need to find out the background of some contractors. Remember, a little legwork can save you a lot of money.

The plumber you contract will need to install the underground plumbing once the foundation walls are up. These installations must follow strict local and national codes and guidelines, so prepare for this during the inspection process. The scrutiny of such installations is in place to ensure the occupant's and the environment's safety.

Water wells are used when there is no access to a town or city water supply. There are stringent Environmental Protection Agency (EPA) rules governing the distance and location of wells and septic systems from each other. Be sure the contractors chosen to dig the well and run the transport pipe protects them from freezing by placing it all below the frost line.

When researching contractors, be sure they are members of the National Ground Water Association (NGWA); this shows their dedication and professionalism. Other designations include Certified Well Driller (CWD), Master Ground Water Contractor (MGWC), and Certified Pump Installer (CPI or CWD/PI). Theses certifications will prove consequential when it comes time for the inspection.

It is a good idea to install water filtration for well water and a municipal supply on the main waterline as it enters the house. Water filters have been proved to remove sediment, lead, dust, sand, silt, sulfur, dirt, excess chlorine, and other contaminants. Filters also reduce the water's odor and can improve the taste. Installation during construction protects pipelines, equipment, and appliances from harmful elements the water may contain. This measure is optional, unless a water treatment tests proves otherwise, so do not worry if you do not have enough room in your budget for this installation. Water filters can be safely and successfully installed later.

When the inspector comes to ensure that all the work meets code, they will also check to be sure all the waste lines are properly sealed. The septic system is usually inspected the same time as the rest of the underground plumbing. If you have a private septic system, it is absolutely essential to hire a reputable contractor to install the waste system so all the codes for your municipality and the EPA are met. Once all the areas of inspection have passed, you or the contractor can fill in all of the trenches.

Remember, plumbing is an important part of the planning process. It is crucial to know what you want and where you want it to be. It is extremely tough and expensive to go back and move lines around or add things in if it is not done right the first time. Be sure to take your time, discuss these plans with others, and ask questions of yourself as well as any other occupants of your house.

Money-Saving Tip:

The importance of proper planning cannot be stressed enough. You will hit highly expensive and time-consuming issues if you do not have a solid set of plans in place. Once the pipes are set and the foundation is poured, there will be no opportunities left to change your mind; be sure to get it right the first time.

8.5: Backfilling, Drainage, and Landscaping

Be sure to inspect the property before any backfilling occurs. Backfilling is simply pushing back excavated soil to fill in the construction ditches. This step normally occurs after the first floor is framed on the house, but it is not a firm rule in the industry. Most contractors choose to frame the first floor to add more weight to the structure and stiffen the walls. Do not be alarmed, though, if you see your excavator moving earth before the first nail is hammered. Make sure there is agreement regarding when the backfilling should occur.

Do not allow your contractor to backfill too quickly; otherwise, the strength of your foundation will be compromised. There are a lot of owner-builders and contractors who will rush this portion of the build so they can hurry up and move on to the next task; this is a mistake. It takes up to 28 days for concrete to cure to 75 percent of its designed strength, so if you are backfilling within a few days of pouring the foundation, you could cause the walls to slant inward.

Notes From the Field

When the backfilling is taking place on the property, check with the excavator about bringing in gravel and straightening out the driveway. The land will be dug up so much that your driveway might soon become a mud pit during the next rainstorm. If this is addressed right away, it will be less of a problem.

A note to remember about backfilling is to make sure the dirt is not compacted around your foundation walls. This will have an affect on water drainage for years or until the earth has completely compacted. In some cases where the overhang is protecting the edge of the ground, the soil does not ever completely compact. Some owner-builders add gravel and crushed rock around the foundation, rather then grassing it there, to help improve any drainage issues that might occur.

When taking into account what type of drainage system you need around your foundation, consult with your foundation contractor to see what they suggest in addition to what the building inspector suggests or requires. They may have good information, and also differing opinions. Be sure these meeting are happening on-site because no one should guess what type of soil and grade you have on your property. Each piece of land is different and should be treated accordingly. If possible, arrange the meeting so both the foundation contractor and the building inspector are present at the same time.

There are many different materials to choose from during the installation process. With a few weeks of planning ahead, just about anything can be brought into your building site and installed. Do not risk quality when attempting to cut costs; otherwise, you could have a nightmare on your hands. These are some materials you will need for drainage installation:

- Drainage pipe
- Geotextile
- Catch basins
- Pipe couplings
- Pipe anchors
- Drainage gravel

Before beginning, if you are unsure of where your utilities are hooked up, call in the utility company service to mark off where all the utilities are located on the property. Some places offer this service for free. Utilities to watch out for include power, telephone, water, gas, and cable. Some municipalities require that these markings be done ahead of time. Be sure to check and make sure before any digging occurs.

The land should be dry before any excavation occurs. The holes dug can quickly become flooded, so it is essential to wait out poor weather and allow the property to dry after storms. Mark out where any natural springs are located so they can be avoided, if possible. This information is available on your zoning map in most cases. Your drainage system will perform much better when installed dry, and there will be no sedimentation occurring in other areas.

Note that sometimes drainage pipes can either break or loosen from heavy equipment. Do not allow the pipes to be buried until they are inspected to ascertain they are not broken and to ensure they will not leak. These inspections are common practice, so the contractor should be helping through this process. If not, request that they do so.

Money-Saving Tip:

Be sure to support the foundation walls with timbers about 12 feet apart if you are unsure if the foundation is 100 percent cured. This will save a lot of disasters, headaches, time, and money.

Not everyone builds their home in the middle of a city or out on a farm. Those occasions frequently call for specialized assistance and expert advice. The materials that work in ordinary situations do not usually fare well on islands, near water, and other wet environments. Building a vacation cottage, for instance, is not such a simple project unless you know the nuances of the area. In general, construction wood and concrete are common materials for building. When building near the sea, they do not do well. We spoke with Arthur Monahan about the problems inexperienced builders face in these locations. It not only serves the interests of sea coast homes, but the information has broad applications.

CASE STUDY: ARTHUR MONAHAN

President, Seacoast Cottage Construction

Ft. Myers Beach, Florida

www.SeacoastCottageCompany.com

"I have witnessed the result of builders who mixed salt water with concrete because fresh water was not readily available. The finished product hid the problem and it crumbled within the year.

But even when used properly, concrete is expensive, difficult to transport, and the steel rebar used can rust over time, requiring costly repair. I have been called onto wood frame jobs to rip everything out and start over because the wood was left exposed to the elements by the original builder, causing it to warp and black mold to set in. However, even if installed properly, wood frame construction lacks strength and is prone to termite damage.

It always amazed me that construction science had not evolved. After some research, I discovered a technology that has been around for more than 60 years and is finally coming into its own: Structural Insulated Panels (SIPs).

SIPs are extremely strong (exceeding hurricane wind requirements) and much more cost effective than concrete. To my delight, they also offer another feature so crucial today: SIPs are the best choice for energy efficiency! With R-values of 24 and higher, I found all the solutions I was looking for:

- Strong
- Cost-effective
- Termite-resistant
- Mold-resistant
- Pre-cut, so less to transport and less waste
- 30 - 50 percent more energy-efficient that traditional construction"

CHAPTER 9
The First Nail
(Phase Two)

9.1 First Things First

Major Considerations for Construction Activities:

1. Be patient.

2. Know about all the materials and tools you will be working with, both from the manufacturer and from a mentor.

3. Never change a blade or adjust a tool without unplugging it first.

4. Wear the proper safety clothing (eyewear, boots, hard hats, etc.) whenever working on the building site. You will need gloves, rubber-soled boots for the roof, hard-soled boots for the ground, etc.

5. Protect your ears when loud equipment is in use.

6. Loose clothing and long hair should be tucked away when operating power tools so nothing gets tangled.

7. If your work site and traffic patterns are as clear and clean as possible, there will be fewer accidents, falls, and damage.

8. Ask for help whenever needed and do not try to attempt difficult or heavy tasks on your own.

9. Take extra precautions when working with or around glass.

10. Even if you believe nothing will happen and you are trying to save time, do not rush through anything.

9.2: Frame Your House

Your house is truly going to start coming together during this phase. The assembling of wood or steel components to form walls, floors, and the roof are all framing. Your framing contractor will spend more time on the building site than any other subcontractor you will work with during the build.

Notes From the Field

Do not act like a know-it-all, as tempting as this may be. If the knowledgeable and more experienced contractor points out something wrong or offers advice about doing something differently, do not argue. Listen to his or her advice, ideas, and opinions and make an informed decision. Do not forget to thank the contractor for the insight. They are there to help you get things right.

The following are descriptions your contractors will use freely and expect you to know:

Floor joists are parallel pieces of wood supporting the sub flooring. Larger beams support the floor joists and, in some cases, bearing walls or girders are also used for support. The measurements for floor joists are normally 2 inches thick by 8 inches wide; 2 inches thick by 10 inches wide; and 2 inches thick by 12 inches wide. If you take actual measurements, you will notice that these boards run a half-inch smaller than what is listed. Do not let this confuse or upset you.

Headers are pieces of wood that are doubled up and used to support floor joists. Headers are also used for windows and doors to allow transfer of the weight of a roof or flooring to the studs. The headers are added to cap off the floor joists by running perpendicular to them.

Floor decking consists of the pieces of wood spread out over the floor joists to create a structured surface for the floor, characteristically constructed from plywood. Wood is laid diagonally across the floor joists. Edges are butted together, staggered over the expanse of the floor. For added durability and less creaking, plywood panels are both nailed and glued to the floor joist.

Bridging and Blocking: Bridging is installed in a crisscross fashion between the floor joists under the sub floor. To work properly, they need to be installed on both sides of the joist. Be sure the wood is placed about a quarter of an inch from each other so there is no rubbing and creaking. You can either purchase the bridging yourself or cut your own. Blocking is installed in a similar manner on both sides of the floor joist in an alternating pattern. Many contractors prefer bridging to blocking because it creates a sturdy structure.

Floor insulation is used because it is less expensive to insulate a floor than it is to build a thicker wall or use sheathing made from foam. Choose a thicker insulation so it is easier to install and will produce a better quality job. This project can be completed by the owner-builder or, if preferred, by a contractor.

Wall framing normally is constructed from 2-by-4 framing, but you can create a sturdier structure by building with 2-by-6 pieces of wood instead. 2-by-6 constructions also allows for more room for insulation. Some contractors also prefer these constructions because there is more allowance for pipes in the bathroom and kitchen areas.

Corner bracing is used to help protect against earthquakes and windstorms. Bracing should occur on all the outside corners of the house and should be placed every 25 feet along the wall.

Openings in wall framing are rough openings framed-in when the wall framing is completed. Measure the rough openings so they accommodate the doors and windows you have chosen for each room. Proper planning for this stage of construction saves both time and material costs.

Ventilation systems are a crucial part of the roughing in phase of the home's construction. Ventilation most expediently occurs when the heating and cooling systems are installed.

Sheathing is the covering of boards, the first layer that is installed to the exterior walls and roof prior to adding the rest of the exterior. This construction element is used to cover roof tresses and exterior walls. The sheathing provides structural support.

Nails come in many varieties and sizes to suit each phase of the carpentry job. It is important to make the right nail choices to avoid sizes that are too short or long. Here is a list of the various types of nails available, along with what they are normally used for:

- **Oval wire nail:** These nails are used for joinery where nails can be seen and appearance is significant.

- **Round wire nail:** An all-purpose nail used where it will not be seen.

- **Round (or lost head) nail:** These nails are also used for joinery and are stronger than oval nails.

- **Tack:** These are used for attaching carpets to floorboards and stretching fabric over wood.

- **Panel pin:** Characteristically used in cabinet construction, this type of fastener is commonly used with wood glue.

- **Cut floor brad:** These are nearly always used for nailing floorboards to joists.

- **Masonry nail:** These are used for attaching wood to most any type of masonry.

- **Square twisted nail:** These are more expensive, but tend to have a better grip than regular nails.

- **Annular nail:** These are used where an exceptionally strong joint is required. In most cases, these nails are used with plywood.

- **Cloat head nail:** Used with soft materials like felt or roofing

- **Spring head roofing nail:** These nails are used for fixing corrugated sheeting to timber.

- **Corrugated fastener:** These are used for fixing weakened areas in roofing.

- **Cut clasp nail:** These nails provide an exceptionally strong joint in wood and pre-drilled masonry.

- **Hardboard nail:** These are best for hardboard because they virtually disappear when hammered into the wood.

- **Sprig:** These are nails without a head that are normally used to hold glass in place before putty is applied.

- **Upholstery nail:** These nails come with a decorative head to add a decorative touch to furnishings. They are used in conjunction with tacks.

- **Staple:** These are used to hold lengths of wire in place.

Money-Saving Tip:

As tempting as it is to build up your arsenal of supplies, the costs associated with doing so will add up quickly. Only buy what you absolutely need. If you are unsure of the specifics, consult with your contractor about the materials list or with a specialist from your local home improvement center. It is important to have enough of what you need, but you will not save money if you purchase unnecessary materials.

9.3: Cap it Off: Roofing

Roofing subcontractors cannot be called in until the framing of the roof is complete, so it is difficult to create a timeline for this portion of the build. A general idea is in the plans and the schedules, of course, but any number of factors could delay this portion of the project. There are many different roof styles to consider when drawing up your house plans:

- **Gable** roofs are triangular in shape with slopes in the same pitch.

- **Gambrel** consists of two slope patterns down each side; the top slope is flatter than the second steeper slope.

- **Mansard** roofs are flat topped with incredibly steep slopes on all four sides.

- **Flat** roofs have a level surface with no slopes, pitches, or peaks.

- **Hip** roofs have no gables and contain four sloping planes all the same pitch.

- **Shed** roofs contain no hips, gables, ridges, or valleys. They consist of one sloping plane.

Gable roofs tend to be the most commonly built in the industry. Frequently, the slopes are each equally pitched at meet at the roof's ridge. These roofs normally create a triangle shape on the side or front of the house. Rakes are used on the gables sides, and eaves are present where there are no gables.

What about the pitch?

Determining roof pitch is calculated by the number of inches it rises vertically for every 12 inches it spans horizontally. Knowing the pitch of the roof is significant for those who plan on making future additions to their homes and is key when it comes time to cut roof rake boards.

Notes From the Field

Do not forget to account for attic space when determining the pitch of your roof. You may short yourself some much-needed space if this detail is overlooked.

What is a roof deck?

The roof deck tends to follow the roof's shape and is located under the roof supports and is what the rest of the roof's materials are attached to. It must be strong enough to support the weight of the materials and have some flexibility. The roofer you contract for this part of the process will be able to help you determine the type of decking materials you will need, depending on the kind of roof you install.

The Importance of Good Ventilation

Roof ventilation allows moisture to leave the attic area, preventing ice build-up on eaves in the winter and an overheated roof in the summer.

When it comes to roof vents, the more the better. A common sign of a roof that is not properly vented and getting too much heat are asphalt shingles that are curling up at the corners. Ventilation is important because it allows air to naturally escape the attic in an upward manner; it promotes a dryer attic, prevents moisture trapping, and lowers energy consumption.

Roof fans should not be installed until the ventilation in the attic is inspected. There are two different types of roof fans: one that is installed during roof construction and one that is installed in the gable wall, which is the area between the edges of a sloping roof.

Chimney vents are significant because though they do not have much to do with the attic's air, they still pass through or next to the attic floor and roof. When performing your routine building inspections, be sure the chimney is positioned at least two inches away from all wood framing. In addition, the chimney should be three feet above the ridge.

The Importance of Good Insulation:

High-quality insulation is one of the best investments you can make in the construction of your new home. Even though it will be more costly for the higher quality materials at the start, in the long run, you will be thankful you spent the money when your utility and energy bills are lower. Products for insulation are rated by their "R-value," which is a measurement of thermal resistance; the higher the R-value, the better the insulation. Because heat rises, the most important areas to insulate in your home are the attic and roof.

Roof exteriors can be constructed of a wide variety of materials, including asphalt shingles, wood shingles, and metal. You must ask yourself, "Which material is good for my new home and what options do I have?" You want to choose what is best, of course, and what will fit your budget. The most common materials found on homes are asphalt and metal shingles. However, there are others to choose from; use this guide to help you decide:

Choosing Materials: Types of Roofing

- **Asphalt** shingles are the most commonly chosen type of exterior. They are made from a base material and covered with granulated particles.

- **Fiberglass** shingles are made with a reinforcing mat, rather than one that is organic. Fiberglass shingles have a good fire rating, are easier to carry around because they are thinner, and tend to be more popular in the South.

- **Slate** shingles require a different installation method, so be prepared. Slate requires a special skill to install and will cost you more for that expertise. Slate shingles are durable and are considered a green building alternative because they are a stone material. They will not sag over time like asphalt, so they carry a lot of added benefit to the home.

- **Wood** shingles are uniform in thickness and resemble shakes. This roofing option is particularly appealing to those who are looking for a simple installation and a rustic look. Homeowners building log cabins tend to gravitate toward this historic look.

- **Tile** shingles serve more as a decorative feature on a roof than anything else. The underlayment provides the waterproofing the roof needs because, as many people do not realize, tile is not waterproof. If this is a project you intend to tackle yourself in order to save installment costs, do your homework first.

- **Metal** shingles are constructed to resemble wood shakes, but they are a better option. Not only do they last longer and are more durable, but they also require less maintenance. Many traditional roofers will not install a metal roof because the material requires specific techniques and, in some cases, special accreditation. Do not let someone who is "winging it" try to install this type of roof for you.

Choosing the color of roofing materials

This can be an overwhelming process. Your roof should complement the rest of the colors chosen for the home. Also, bear in mind how the outdoor temperature will affect the color. Darker colors have been known to absorb and retain heat more than lighter colors.

What is Flashing?

Flashing is used for weatherproofing around areas of the roof where there are protrusions, such as the chimney and pipes. It is also used on the doors and windows where they protrude from the wall. The flashing is meant to divert water away from the seams. Without flashing, there is nothing to prevent water seepage. Never substitute caulking for flashing; this is a common mistake around chimneys.

Water drainage: For most homes in the United States, it is necessary to install a drainage system that will collect and draw away rainwater from the roof. This water drainage prevents water from getting into basements and crawlspaces, and splashing against the siding of the house. It also prevents the soil from eroding away from the foundation.

Money-Saving Tip:

Despite the initial high cost associated with metal roofing, these roofs have proved to be a wise long-term investment. They require less maintenance, are extremely durable, have longer warranties (depending on the manufacturer), and reduce energy bills due to their reflective properties. In addition, some homeowners experience lower homeowner's insurance rates because metal roofing is non-combustible, weather resistant, and is considered among the safest choices in roofing materials.

9.4: Windows and Exterior Doors

The final step for closing in the house is installing the windows and the doors. The framing contractor normally does this, and it should only take a day or two to complete. If you are an experienced do-it-yourself-er and you choose to install the door on your own, you can do so by following the following steps:

1. Frame up the rough opening.

2. Be sure the rough opening is plumb, and that the sub sill is level.

3. The rough opening should be at least one inch larger than the frame of the door.

4. Add a generous bead of caulk along the sill, about one inch from each edge, and up the sides.

5. Place the door in the opening bottom first.

6. Center the door into the opening, then add shims to keep it tightly in place.

7. Add shims next to hinges, adjusting them until they are plumb in both directions.

8. Once the door is plumb, drive 16 finishing nails into the hinges. Do not hammer them all the way in.

9. Remove all the shipping assembly, and then test the door to be sure it opens and closes freely.

10. On the outside of the door, make any adjustments necessary to ensure the door is level with weather stripping.

11. Install a solid shim behind the lock strike location of the door on the inside.

12. Permanently screw in the door using 3-inch screws, starting at the hinges.

13. Add insulation around all the gaps around the door frame. It is not advised that foam insulation be used because it could distort the frame of the doorway as it expands.

14. Attach interior molding as per the manufacturer's guidelines.

15. Install weather stripping at the base of the doorway as per the manufacturer's guidelines.

The U.S. Department of Energy has reported that as much as 25 percent of your heating and cooling bill is affected by heat gain and loss through exterior doors. As you can see, it is important to follow proper installation guidelines and also install a storm door over existing exterior doors. Energy efficiency is increased by nearly 45 percent with the installation of a storm door. If it is not in your budget to do so now, add it to your list of additions to make in the near future.

The same is true if you would like to install the windows of your home on your own. Because you may be working at heights requiring the use of a ladder, be sure to enlist of the help of friends or family members. Not only will this ensure your safety during installation, it will speed up the process and prevent the threat of breakage.

When ordering your window, you will need to know the depth of the wall for it to fit properly. Pre-hung windows come with just about everything you need for assembly and installation, including a frame, sill, installation hardware, and some of the trim work. For better fit and safety, installation from the outside of the home is recommended, as well as having someone who is capable of assisting.

Notes From the Field

You may be surprised where cost-effective windows can be purchased. Look for them in small Mom-and-Pop stores. More often than not, these out-of-the-way retailers are less costly than the more convenient home improvement stores. They are generally quite knowledgeable, are a good resource, and a lot more personable. If you come in with a sales flier from a home improvement center, they may even meet or beat the prices listed. It never hurts to ask, especially when it can save you even more money.

Follow these steps for installing your storm windows:

1. Create a rough opening for the window, allowing room for the brick molding or flange around the perimeter of the window.

2. Apply 8-inch wide moisture stripping around the perimeter of the opening for the window.

3. Staple the stripping into the corners.

4. Add shims to the base of the rough opening, adjusting them until they are level.

5. Drill pilot holes to prevent the shims from splitting.

6. Nail the shims into place using 6d 2-inch nails, then cut them flush with the exterior wall.

7. On the inside of the house, trim the shims so they are flush with the window jamb.

8. Center the window in the opening and, while holding the window in place, nail one of the corners into place. Be sure the window is still level.

9. If you are using brick molding, nail it in place using galvanized casting nails (8d), and, if you are using a flange, use 4d-roofing nails to nail it in place.

10. Thoroughly caulk all the joints. There should be manufacturer's guidelines to follow in terms of flashing and caulking around the window trim.

11. Finish the inside of the window with molding.

In addition to reducing energy bills, the installation of storm windows also helps with noise reduction. If this is not in your budget right now, it is still possible to add storm windows later. There are interior storm windows you can consider adding that have the same benefits, but they do not alter the exterior appearance the home. Look at each option carefully before making your final decision.

Money-Saving Tip:

Purchase Energy Star windows. They are double-paned thermal windows that will give you a rapid return on your investment in energy savings. Most home improvement stores have brochures and other related literature for consumers to take home with them and learn about the many cost-saving benefits this choice in windows has. If you cannot find these materials at the home improvement stores, do a search using your favorite search engine to produce similar materials online.

9.5: Decks and Porches

Depending on what you have planned, it is now time for the decks and porches to be built. The framing contractor will create a rough deck or porch, but commonly does not complete the finished addition. If you do not know how to do this on your own, it will be necessary to bring in a separate contractor to complete this task, or spend some time browsing the do-it-yourself section of a home improvement store. There are many books available that are devoted to building porches and decks. The guide below should be sufficient to get the job done, and a book is probably not necessary. The job is labor-intensive but not difficult if the frame is already laid.

A lot of do-it-yourself-ers love building their own decks and porches. This does not have to be a complicated project because, ultimately, the goals as of now are to save money and pass inspection. In some municipalities, the rough deck is enough to pass inspection. Check this out ahead of time to be sure. It will have an effect on your budget, so it cannot be overlooked.

The first step is to come up with a basic design for your deck. Once that is complete, you will be able to create a supply list and gather all the materials needed to complete the project. The most common wood choice for both decks and porches is pressure-treated wood. Redwood and cedar are used — based on budget — sometimes for aesthetics. The overall structure, though, is normally constructed from pressure-treated wood because it is durable and holds up to weather.

Notes From the Field

For those without a light-duty truck, you may have the option of renting one from your building material supplier to haul materials or have them delivered. Inquire about each scenario, then consult your budget to see which option is best. Often, having the materials delivered is the most efficient. It saves you the time and effort of loading and unloading, especially if you have no help.

The next step is preparing the site for the work. Remove all the sod from the ground surrounding the area and grade it away from the house for proper drainage. From here, follow these basic steps:

1.   Measure the wall for the ledger, about one inch below the doorway, plus the thickness of the decking.

2.   Mount a 2-by-6 ledger using half-inch lag screws. Be sure to keep it level and that the screws go into the stud at least three inches.

3.   Screw in two lag screws on each end, then in a "Z" formation on every stud in between.

4.   Establish the sides of the deck by running string taught from the edges of the ledger and out to batter boards. Set the batter boards into the ground. Make sure the layout is square.

5.   Measure for the footings, which must be at least 6 inches below the frost line. Fill the bottom of the hole for the footing with gravel for drainage.

6.   Mix the concrete, then pour the footings. After pouring each footing, set in a pre-cast pier so it sits 6 inches above the ground.

7.   After the concrete has set, set the posts on the piers. Use temporary braces to hold the posts in place. Once they are braced, run a line from the ledger to measure and mark for cutting.

8.   Fasten the posts to the beams using nails and hex bolts. Make sure everything is plumb and level.

9.   Mark the joist locations on both the ledger and the beams. Set the joists with the crowns up. Determine if local building codes require blocking between joists.

10. Build staircase as needed.

11. Center the deck boards over the joists, and then nail them down so they are staggered and do not line up.

12. Let the boards run off the edge of the deck, then saw them off.

13. Secure railing posts at each corner of the deck and on each side of the stairway.

14. Nail the railing in place, then add balusters.

Not only do decks and porches provide an additional area for family seating and entertainment, but they also provide an additional opportunity to create an outdoor room. Space planning is an essential part of the preparation process. What do you intend to use the deck or porch for? Is it merely a functional area constructed to enter and exit the house? Would you rather have an area where furniture and other items can be used? Take all these thoughts and plans into consideration before taking measurements and making building plans.

Money-Saving Tip:

An alternative to pressure-treated wood is composite decking materials. Homeowners have been making this material choice for years because it stands up to the elements and require extremely little maintenance. Pressure-treated wood requires regular applications of stain or weatherproofing to keep away rot and prevent destruction by the sun. Composite requires none of the routine protection and cleans easily with elbow grease or a power washer. Despite the initial up-front cost, in the long run, this saves the homeowner money.

9.6: Siding and Trim

You have reached the final step weatherproofing the exterior of your home. Siding materials include painted or stained wood, vinyl, aluminum, stucco, brick, or stone. Trimming regularly consists of vinyl or aluminum, and is used to cover exposed areas that the siding does not cover.

Vinyl is an affordable choice and can be completed by the owner-builder. Follow these steps for installing your vinyl siding:

1. Start with the trim. Trim serves two purposes, either to hide edges or to hold everything together. Trim pieces have the same names as wood trim pieces. These include rakes, soffits, fascia, and corner boards. Start by installing the high trim first.

2. Keep corner boards straight by snapping a chalk line and using a level. It can be difficult to hang a corner board straight, so snapping a line on each side will help ensure it is. Start nailing from the top.

3. Use tin snips to cut the flange back where it butts up against the trim.

4. Allow for expansion by overlapping vinyl pieces so they can expand and contract with temperature changes without buckling. If the nails are not driven in all the way, it allows for the vinyl to move sideways without causing damage.

5. Plan the layout of the vinyl runs. There are three different kinds of vinyl patterns to choose from: double four, triple three, and double five. Once you have chosen the pattern of the vinyl, determining how it will run will be more efficient.

Notes From the Field

Keep all the large scraps and leftover pieces of siding once the installation process is complete. Store them in a place where they will not get damaged so they can be used for future maintenance projects, additions, and repairs if needed.

If vinyl siding does not interest you, there are a number of different options you can choose from:

- **Aluminum siding:** This option is appealing to those who are looking for something that is inexpensive and durable. However, it dents easily, so maintaining the exterior of your home can become a challenge. Environmentally conscious consumers have considered this type of siding to be a somewhat conservative material.

- **Brick siding or brick veneer siding:** If you like the look of brick but the cost is out of your budget guidelines, consider the installation of brick veneer. It lasts about 25 years before anything needs to be done, so it is a cost-saving investment in the long run. In addition, there are a wide variety of colors available outside of the traditional red brick.

- **Cedar shingle siding:** Homeowners interested in natural or rustic looks tend to choose this type of siding above all others. Normally, the shingles (or shakes) are stained rather than painted (like clapboards). This results in less maintenance and upkeep, which is cost-effective.

- **Engineered or composite wood siding:** If you are looking for a cost-effective and easily installed siding option, then engineered or composite wood is your best bet. These pieces of siding are available in long lengths, providing a seamless look to your exterior. This is becoming a highly popular choice when homeowners are looking for an inexpensive alternative.

- **Seamless steel siding:** Durability, safety, and easy maintenance are just some of the words used to describe this siding type. In addition, there is no need to paint, and this siding offers the same seamless look as composites. Steel can also be manufactured to resemble wood for those who are looking for the best of both worlds.

- **Stone siding:** This tends to be the most expensive choice but is also the most durable. It has beautiful curb appeal and is considerably difficult to damage. Stone veneer and pre-cast stone options are available for less of an investment.

- **Stucco siding:** This type of siding creates a kind of shell around your home that provides strength and durability. Normally, it is constructed of cement and other natural ingredients. This truly hard, like-a-rock surface is extremely durable and moisture resistant.

- **Wood clapboard siding:** This type of wood, when properly maintained, will stand the test of time; look at some century-old homes as a testimony of this statement. There is a wide variety of wood that can be chosen for this sort of siding, including pine, spruce, fir, and redwood.

Money-Saving Tip:

Many owner-builders choose vinyl siding not only because it is cost-effective right from the beginning, but because it is a material that is easy to maintain. Homeowners save money because they do not have to paint, nor do they have to worry about other expensive types of upkeep or cleaning. Cleaning siding is as easy as using a mild detergent and hosing it down periodically. Vinyl siding is also durable.

CHAPTER 10
It Looks Like a House! (Phase Three)

10.1: Electrical and Plumbing

Roughing in the plumbing, depending on your town or municipality, might have to be completed by a licensed plumber or electrician. There is nothing temporary about roughed-in work. Do-it-yourselfers can handle most projects so long as they are cautious. If you do not feel confident enough to handle something that is intimidating to the untrained eye, hire contractors.

Before roughing in the electrical, double-check all the diagrams. It may seem like an easy enough process to run cables and wires through the framing, but you will find it requires a lot of patience because the wires may not run the way you expect. With time and creativity, it will work out. Remember that before the walls can be hung, inspections must take place.

Notes From the Field

Call the inspector well in advance to give him or her plenty of lead-time to reach the building site. Communicate this to your contractors so they can give you a ballpark estimated date of completion. One owner-builder reported that his failure to do so threw off his schedule during the rest of the building process and ended up costing him extra money in manpower.

Roughing in the plumbing consists mostly of the main water supply and drainage lines. Be sure you have all the proper tools and materials in place before you begin. Your home will depend heavily on how efficiently the drainage, waste, and supply lines operate so it is exceedingly vital to find outside help if you are unsure about anything to do with the installation process. Some tools needed for roughing in plumbing include:

- **Pipe cutters:** This tool allows you to cut put at lengths needed.

- **Pipe threaders:** This threads pipes after they have been cut.

- **Caulking irons:** This is used to distribute caulking material.

- **Chisels:** The sharp beveled edge allows for shaping of wood, metal, and stone.

- **Saws:** These will be used both for pipe and for wood. The same saw does not cut both.

- **Wrenches:** The pipe wrench has two toothed jaws: one that is free to move and one that is fixed.

- **Files:** These are commonly made from steel and used to smooth the rough edges after a pipe has been cut.

- **Plumb bob:** This is also referred to as a plummet, the metal weight that is attached to the end of a plum line.

- **Machinist's hammer:** This is a hammer with a square head, beveled edges, and beveled corners.

- **Level:** This tool will ensure everything is in line horizontally, but it also works on the vertical axis.

- **Square:** This will ensure all right angles are met.

- **Ruler:** Preferably wooden, to ensure safety if contact with anything electrical is accidentally made.

Be sure there are enough bracing supports installed to account for the pipe's weight empty, and also when water and waste pass through. Inadequate pipe bracing causes sagging and leakage, plus the loosening of joists. To prevent trouble, brace both the top and the bottom of pipes with a sturdy brace or clamp no more than 10 feet apart from each other.

Roughing in the electrical consists of nailing up circuit, outlet, and switch boxes; calculating box sizes; drilling holes; and connecting wires. Any experienced do-it-yourself-er can accomplish these basic wiring techniques. This guide is meant to act as a way to get you started. If you have more complex wiring issues to contend with, hire an electrician. Also, remember that a lot of homeowner's insurance policies do not cover homes unless they are professionally wired. Check this out before getting started. If you are able to complete this task on your own, here are some basic guidelines:

1. Plan out where the boxes will be located. Mark where each light switch and electrical outlet will be installed on the wall studs. Measurements should be taken from the floor to the center of the box (48 inches for light switches and 12 inches for electrical

outlets). Nail the box in so it is flush where the drywall is going to be installed.

2. Consult with an electrician or an experienced electrical do-it-yourselfer if you are unsure how to calculate the electrical limits of each box size.

3. Drill the holes and pull the cables through. Again, there are specific guidelines for measurements that must be followed to ensure safety. It is imperative to learn this information first-hand from an electrician.

4. Install electrical outlets in all walls that are two feet wide or larger.

5. Connect all the wires. Many owner-builders will work right up to this point, then call in an electrician to inspect the work completed and make the final connections. Be aware, though, that a lot of electricians may not undertake this task because of liability issues; they can be hesitant to put their name on a project someone else performs.

To take on this project on your own, you will need some basic materials. Unless special circumstance arise, you will need nonmetallic cable, plastic boxes, wire connectors, a bag of 50½-inch staples, a half-dozen metal nail plates, and a roll of black electrical tape to mark white wires.

Money-Saving Tip:
The cost of fixing electrical and plumbing issues is huge, so making mistakes could complicate or break your budget. If for any reason you feel you are not experienced or qualified enough to take on these projects, do not attempt to complete them on your own. The costs associated with fixing your mistakes can have a very negative impact on your budget, if not your ego.

10.2: Insulate

Before insulation can occur in the walls, the ductwork, pipes, and wires need to be installed and inspected. Insulation occurs both on interior and exterior walls. An interior wall is one that does not face the outside. Because construction is so standardized, insulation of one interior wall will be the same as insulation of another interior wall in any other room.

Heat rises, and one of the biggest heat transfer sites is through the ceiling. This is particularly true for single-story homes. If the attic is not insulated — or improperly insulated — heat in the home will quickly rise, move through the ceiling, and into the attic space. From there, it will escape through the roof. Heat loss through the attic, of course, will make the interior living areas of the house uncomfortably cold during winter months.

This process works in the reverse during summer months. Attics can be as hot as 160 degrees during hot summer days. The heat from the attic will work its way down into the interior living spaces, causing temperatures to rise and requiring additional air conditioning and fan use. Proper insulation can reduce heating and cooling costs by as much as 30 percent per year. When insulating the attic, go with then next highest R-value recommended for your area.

Window energy conservation should be addressed as readily as wall and attic insulation. Caulking and sealing gaps is a crucial part of the insulation process. In addition to windows, this also includes around doors, vent openings, ceiling fixtures, and framing gaps. Insulation of the foundation, floors, and basement is also key in terms of energy reduction. If insulation is installed correctly, it is impossible to use it in too many areas of the home. Follow the building codes and the manufacturer's guidelines for installation, clearance guidelines, and other important requirements.

Notes From the Field

Old-school carpenters will offer their input about how you are wasting time and money insulating where they feel it is unnecessary. Listen to them respectfully, thank them for their advice, and then do your own research. If you are at all doubtful, consult with an insulation contractor or visit a reputable Web site to research this further online. Glen Haegy's radio show or Web site (www.masterhandyman.com) is one example. Information should come from individuals with a reputation of reliable advice.

Insulate heating ducts, cooling ducts, pipes, and the hot water tank. Taking these measures will reduce energy costs up to 10 percent per year. Be aware, though, that some hot water tanks do not allow for insulation to be placed on them. Read the manufacturer's label to ensure doing so does not void the hot water heater's warranty. Also, be sure any bathtubs and showers installed on outside walls are heavily insulated. Builders often overlook this process, which causes energy loss as a result.

Money-Saving Tip:

Read the labels. Depending on how you are heating your home, the R-value of the insulation will differ for what is installed in the attic space. Walls should be insulated with an R-value of 19, and the upper portion of the basement (the sill box) should be insulated with an R-value of 10. Following these guidelines will save on energy costs.

10.3: Drywall and Trim

Covering the interior walls of a home is usually done with drywall, which is also referred to as gypsum board or wallboard. The steps for installing drywall are relatively easy, although the heavy boards can be awkward to handle. Here is what you will be doing at a glance:

1. **Hanging the drywall:** Always work from the top to the bottom, hanging sheets perpendicular to the studs.

2. **Taping the seams:** This process takes a lot of patience and practice.

3. **Applying joint compound:** Keep the compound at 1/8-inch thick with a joint knife.

4. **Sanding:** This is a lengthy task that also requires fortitude.

5. **Priming:** Priming uses the same techniques as painting with brushes and rollers. It may be necessary to apply more than one coat of primer for even coverage. For the best results, allow coats to dry completely between applications.

If you can swing a hammer (or turn a screw), you can install drywall. Even though the process is relatively simple, it is still necessary to employ some extra hands. The pieces of drywall are rather large, and can be heavy. If the material is too heavy for one person to work with, it could break or be hung crooked. The best course of action is to have one person hold the piece of drywall while the other person secures it to the studs. If you are unable to find anyone to help you with this project, it is not impossible to achieve this goal alone. Simply cut the pieces of drywall in half. Here are some tools you will need when installing the drywall:

- tape measure
- drywall saw
- utility knife
- wall joint compound
- joint tape

- tin snipe
- drywall nails or screws
- hammer or electric drill
- metal corner protectors
- primer

Note that screws are the preferred method for attaching drywall to studs. They are easy to remove if, when installing, you do not hit a stud. Use of a stud finder will prevent this from happening in nearly all cases. Also, remember that studs tend to be spaced 16 inches apart in most framing scenarios.

Notes From the Field

Some contractors have reported that the learning curve for drywall installation is a tricky one. This is due in part to the taping of seams and the sanding of joint compound. You can do it as long as you do not get frustrated easily.

Measure and cut the drywall to fit the area where it is intended to be hung, or hang the entire piece if there are not obstructions that need to be cut around. Start screwing — or hammering — in the drywall to the studs, starting from the middle of the piece and working your way out. When hanging the bottom piece, allow for expansion by leaving a gap between the piece of drywall and the floor; this will be covered with trim later. Measure, mark out, and cut all obstructions such as electrical outlets, windows, and doors, prior to hanging the board.

Sharp, tight corners can be achieved through use of the metal corner protectors. Bring the pieces of drywall together in the corner so they are flush with one another. Then, nail or screw the corner protector in place. Installation of this material is vital for achieving this look. This process will also prevent nicking and other damage during the finishing phase.

When the installation is completed, cover all the screws or nails with joint compound, and tape all the seams. Then, apply joint compound over the taped areas. The material must dry completely before any sanding can occur. It will be necessary to reapply more joint compound after sanding. This process will be repeated three times, or until the wall is smooth. After this, the surface is ready for primer.

The trim, which is also referred to as molding, is installed after the drywall has been completely finished. This is often referred to as finish carpentry, and also includes the installation of interior doors, door trim, windowsills, window trim, baseboard trim and, sometimes, closet shelving.

Interior molding serves three basic functions:

1. Ceiling molding is used to cover the small gaps between the wall and the ceiling.

2. Floor molding is used to cover the small gaps between the wall and the flooring.

3. Door and window moldings are used to cover the small gaps between the walls and the doors and the windows.

See the pattern here? The primary uses, aside from serving decorative needs, for molding are gap coverage. There are many other uses for molding, but this is the most common. Once you are able to master this type of installation, any other kind of trim work can be achieved successfully.

Molding does not have to be made from wood. There are many other materials that can be used. To prevent splintering and splitting, pine and spruce are recommended most often. This is because they are softer woods and are inexpensive. To prevent warping, finger jointed (FJ) molding is recommended. Rather than being one solid strip, it is molding that is attached end to end. It also cuts down on costs.

If you choose hardwood options and are not a wood craftsman, it is best to have a professional install it, as these woods tend to be tricky to work with. Reproductions are available, but they should also be professionally installed due to the same difficulty they tend to possess when being worked with. Because of the costs associated with higher-quality wood materials, the last thing you want is a pile of cutting mistakes sitting on the ground, rather than finished molding in your home.

Pre-finished moldings are available with stain, veneer, or varnish. The veneer finishes are cost-efficient, but less resistant to moisture. These finishes should not be used in areas of high-moisture potential, such as kitchens and bathrooms. The materials are characteristically made from particle-board with a vinyl covering that has a wood-grained pattern.

There are many types of molding. The most common are as follows:

- **Crown molding:** This type of molding is placed along the ceiling to create a decorative, triangulated edge.

- **Case molding:** This is the molding found around doors and windows.

- **Bed molding:** This is another type of decorative ceiling molding.

The tools you will need to complete this project include:

- Molding
- Hammer
- Nail set
- Backsaw
- Nails
- Steel measuring tape
- Level
- Miter box
- Coping saw
- Putty knife
- Wood putty

Installation should start with what carpenters call "standing trim," which is found around door and window casings. Be sure the casings are flush before starting this project. If they protrude, it will be necessary to plane them down. If they are recessed, it will be necessary to add a small piece of material to bring the casing flush to the surface. And if you are unsure of any of the steps necessary to complete this process, seek outside help from a professional or from someone you know who has done this kind of work. However, have confidence in yourself if you think you can do it.

Remember that a wider molding along the ceiling will make the room look lower, unless you are working with a cathedral ceiling or one that is exceptionally high. Use a more narrow molding to avoid that closed-in look. Ceiling and other moldings can be used in conjunction with each other to complete an architectural look.

Running molding is the base and ceiling moldings. When purchasing strips of molding, try to buy them long enough so you can install them without joints. Locate the studs in the wall, and mark them lightly with a pencil. To install the base molding, it is necessary to measure from the door to the nearest wall. Cut the ends of the molding square, then nail them into to studs using 6d-finishing nails. Continue measuring pieces and, using a coping saw, cut them to size and attach them to the wall. Do not forget the old carpenter rule: measure twice, cut once.

Ceiling molding is installed in much the same way as base molding. The difference is it does not attach to the wall, but rather it is installed at a 45-degree angle at the joint. To avoid damaging any of the molding, nail it in one-eighth of the way in, then use a nail setter to finish it off.

Notes From the Field

If you do not use the correct size nails, the ceiling molding will not stay in place. Owner-builders who have made this mistake in the past were faced with sections of molding popping off in random areas of their home. The cost in time and damage was detrimental to their budget

Money-Saving Tip:

Lower-grade wood can be used if you plan to paint the molding. Additionally, you can purchase unfinished material and stain it darker to resemble a more expensive wood species.

10.4: Finish Off the Kitchen

The time has now come to finish off the kitchen. We will break down, in simple chunks of information, what needs to be addressed in order to conclude your kitchen work, in preparation for your final home inspection by the code enforcement agent. Keep in mind that there are a number of design considerations that might be better addressed following the completion of what needs to be done in order to pass inspection; however, this is dependent on your budget and your timeline.

Appliances: When selecting appliances, it is best to choose the most energy-efficient ones that will provide the most benefit for your family's (or your growing family's) needs. If you decide to cut costs with these choices, you will regret this decision in the long run when costly maintenance, repairs, and replacements are in order. Choose what your budget allows for, of course, but do not cut corners here. It is better to buy a high-end model with a slight dent or other non-functional damage at a freight liquidator than an off-brand or poor warranty item on sale.

Cabinets: Once walls are primed in the kitchen, cabinet installation can take place. Be sure to remove cabinet doors and drawers prior to installation to avoid dings and dents during the process. Also, prior to installation, lightly mark where the studs are positioned in the wall. Start with the wall cabinets first, working from the corners out.

Counter top choices: These choices range from laminate to stone to granite and everything in between. Choose a material that is easily maintained, matches the rest of the kitchen's design, and will hold up long-term. If you are installing the counter top now and a backsplash later, the backsplash will cover any gaps between the wall and the counter top.

Dining area: Creating a dining space can be something as simple as allotting for a small table and a few chairs, or the creation of a separate room with multiple pieces of furniture. Whatever the case may be, it should be large enough to accommodate your family's needs. It should also account for any space limitations you might be facing during the building process.

Lighting: Lighting is often overlooked during the planning and building phase of a new home because, characteristically, these plans and the build occur when natural light is flooding the room. Even if general lighting is addressed, it still might not be enough in the work areas of the kitchen at nighttime. Installations of under-the-cabinet lighting will solve this problem and give the kitchen a more finished appearance. Task lighting can also be achieved through use of recessed fixtures, eyeball lights, and track lighting.

Notes From the Field

It is better to have too many lighting options in your plan than not enough. To account for possible cuts, go overboard with your initial lighting plan as a way of padding for this possibility.

Ventilation: To determine the kind of ventilation needed for your kitchen, the following factors must be addressed:

- Type of cooking equipment
- Type of hood needed
- Type of blower required
- Hood height needed
- Hood width needed
- Hood depth needed
- Type of ductwork needed

These factors are based on local building codes, the kitchen's layout, and structure. Personal preferences, budget, noise control, and airflow also play key roles during the decision-making process. Design features in ventilation hoods range from modern to traditional and everything in between, so be prepared to have these choices affect the decision-making process. If you are prepared, these decisions will not cause a delay.

Finishes: Wall finishes are any material used to cover, protect, and decorate walls and ceilings. Choose something that will stand up to repeated washing and is resistant to staining. If you choose tiles, remember that their durability might be at risk over time in terms of chipping and grout maintenance.

During the planning process, you will be working out a lot of the kinks associated with this finishing-off process. There should be no hard decision-making during this phase of the construction process (unless something goes wrong or a wrong shipment is received). Rather, this phase should be about getting the room finished and ready for the final inspection.

Money-Saving Tip:
Choose appliances that fit your family's needs. If you choose one that is too big, energy is wasted and bills climb. Choosing Energy Star appliances can also help bring energy costs down.

10.5: Finish Off the Bathroom

Types of bathroom sinks range from pedestal to vanity to vessel. There are also console table sinks, which are supported by either brackets or decorative legs. You may also want to consider a wall-mounted sink, which is supported directly by the wall. Space constraints, style, and functionality all play a role in choosing the best sinks for your bathrooms.

There are many common mistakes made during bathroom sink installation, including:

- Violation of local building codes
- Wrong pipe size
- The wrong fittings
- Avoidance of the proper tapes and compounds
- Making the wrong cuts at the wall
- Improper alignments
- Failure to bleed lines by flushing the toilet or running an outside water hose

To install the sink, you will need stub outs for hot and cold water, shut-off valves, flexible tubing, and transition fittings. If you are unsure of any of the methods and guidelines used during sink installation, ask at your local hardware store, home improvement plumbing department, or hire a professional plumber if necessary. There are numerous guides available to walk you through this process accurately, efficiently, and successfully. However, any experienced do-it-yourself-er should be able to handle the installation of a commercial showerhead.

Mistakes made during bathtub and shower installation mirror that of sink installation point-by-point, so be aware of what you are doing at all times. As noted in the section of this book about insulation, be sure to properly insulate behind the shower if it is being installed on an outside wall. Failure to do so will result in energy loss and drafty showers.

To install a bathtub and shower, you will need a pipe leading to the showerhead and stub outs for the hot and cold water. You may also need an air chamber and a mixing valve for the showerhead. The wall should be open for proper showerhead and faucet assembly.

Notes From the Field

Be very selective when it comes time to pick out your bathtub. If you have room in your budget, then choose something extravagant. It is a lot easier to get what you want the first time around than it is to do a bathroom renovation later on.

Some faucets are exceptionally easy to install, while others can be tricky. Plumbing access has a lot to do with how easy or how much trouble you will have with this phase. The best way to avoid problems during installation is to assemble everything before putting the sink or bathtub in; this will also save time. Basin wrenches will help the process along, but it can still be an awkward process.

Shower nozzles come in many varieties, both simple and luxurious. Some popular choices are:

- Rain showers
- Free-hand shower heads
- Star-shaped shower heads
- Square-shaped shower heads
- Waterfall showers
- Commercial (highly durable) shower heads
- Trio shower heads
- Pivoting shower heads

Installation of shower nozzles should take place at the time the tub or shower enclosure is installed. Depending on the style choice you make, installation could be complicated and may require hiring a professional. Some common mistakes include:

- Ignoring or not adhering to building codes properly
- Using the wrong size pipe (too big or too small)
- Neglecting to tape threaded joints
- Cutting stub outs incorrectly
- Failing to add an air gap filling for fixtures
- Failing to level fixtures during installation
- Failure to bleed the pipes by turning on outside water fixtures and flushing toilets

There are several key factors to consider when selecting the right bathroom cabinets. Take into consideration the design and style of the bathroom. You need to decide what type of cabinets or vanity is right for your budget, where they will be bought, and who will install them. Function plays the biggest role over color and texture, but all should be considered when making these choices. Because these fixtures are permanent and should match the rest of the color scheme, take your time during the decision-making process because you will live with it for years to come.

Ventilating fans in the bathroom are a critical part of a home's design and construction. These fans remove both odor and moisture from the bathroom. Without proper use of a fan, mold and mildew could grow from the elevated humidity in the room and also cause damage to wall coverings over time in the way of blistering paint and bubbling wallpaper. If install a fan that is too small, you may be faced with these same issues.

Choosing a fan that is energy-efficient and quiet is beneficial. The size of the unit depends on the size of the room, so check manufacturer guidelines. Be sure to choose a fan that is suitable for continued use and that gives you the ability to replace parts as needed. Fans should be positioned as close to the shower area as possible for the best results.

Bathroom lighting tends to fall low on the list of priorities, especially when it comes to cutting things out of the budget. Introduction of general lighting is easy, but with the addition of some recessed lighting, the room becomes calming when larger tubs or steam showers are in use. Add lighting over the mirror that will not cast shadows on your face. The addition of a swiveled-out lighted mirror can also be useful if you have room in your budget for this decorative and functional option. A popular item to consider is a heat lamp for those cold days after a hot shower.

Money-Saving Tip:

Install low-flow fixtures and water saving devices. This will reduce hot water use and bring energy bills down. Read the literature carefully to avoid cutting down the pressure.

CHAPTER 11
Working the Walls

11.1: Flooring

Before any cabinets, vanities, or toilets can be installed, flooring must be in place. Many flooring options are available, including vinyl, hardwood, and tile. Visit a showroom to determine what style, grade, and color you would like to use in your home. Choosing the right flooring for your home is a big decision, and the flooring should fit the lifestyles of everyone living in the home. These choices set the tone for the design of the entire home and create a foundation for whatever will be placed in the rooms.

If you choose light-traffic carpet, remember it is extremely high-maintenance. Regular vacuuming and shampooing is a must, and staining could become an issue. The color choices are limitless, making for an especially attractive room addition. It will be a lot of work. However, if your budget limits you to this choice, be sure to add its care regularly to your home maintenance plan to increase its lifespan. If the carpet comes with specific cleaning and care instructions, store them with your home maintenance plan to be sure stains and other soiling is handled correctly.

A solid surface, such as hardwood or laminate, is easier to maintain and offers an attractive finished look. These floors clean up easily and are incredibly durable. Solid surfaces are increasing in popularity, particularly for larger families or homes where there is a lot of high traffic areas. Be aware of scratching issues, however, with lower quality laminate floors. This is less of an issue in darker rooms with less window glare.

When making flooring choices, you may be tempted to find the most rock bottom prices available in the marketplace. If you are choosing to use hardwood flooring, take note that the lower-end pieces of wood are the cheapest. Not only will the quality of the wood be compromised, but the look of the material will also be an issue. To some, drawbacks like this do not truly matter. On the other hand, if you are thinking about the long-term investment, it might be worth it to upgrade to a better quality of wood. The higher quality of wood will increase the value of your home if you intend to sell it in the future sometime.

Installation depends on the flooring choice you make. In some cases, like with a linoleum floor, a do-it-yourself-er could handle the project. Conversely, with more complicated work like hardwood or tiling, a professional may have to be enlisted. There are a considerable number of intricate cuts required for these types of installations, and while it may seem easy to watch, it might not be that easy for you to complete.

Notes From the Field

Try to choose the highest quality flooring material you can afford. You will quickly notice how lower quality flooring is not as durable and does not hold up well to high traffic.

Money-Saving Tip:

Do not be afraid to check out the lumber liquidators that have spread through the country over the past decade. They receive high-quality flooring that has been discontinued by manufacturers, over-purchased by complex contractors, and from resellers who have gone bankrupt. You may be saving thousands of dollars.

Remember, no other surface in your home will receive as much use as your floor. While it is rarely thought of while people are walking on it or enjoying a movie in a comfy chair, there are vital considerations to make in terms of durability, longevity, maintenance, and design features. Did you make the right flooring decision? Ask yourself the following questions:

1. What are the primary functions of the space?

2. Is this flooring choice easy to repair?

3. How much traffic can we expect this flooring option to receive over time? Does the material account for high-traffic areas of the home?

4. Can this flooring option be easily maintained?

5. Does this flooring option have a long lifespan? What will compromise the lifespan of this flooring option?

6. Will you be able to recycle the material this flooring option is made of if it is changed out?

Money-Saving Tip:

If you choose laminate flooring, it is easy enough to install yourself. Gather up some friends or family members to help you along with the process. If you opt to have it installed professionally, costs are still relatively lower than installation of a hardwood floor.

11.2: Kitchen Cabinets and Counterpoise

Now that the flooring is installed, it is time for the kitchen cabinets and counter tops. With so many options available, this phase of the design and construction is challenging to some. Your dream notebook will come in handy during this phase of the process.

Notes From the Field

You can collect a lot of materials from home improvement stores containing samples and examples. In addition, visiting home shows and model homes will give you a bird's eye preview you can photograph. Collect all this information to help narrow down your choices.

Here is a breakdown of popular counter top choices:

Granite

This material is an expensive, luxurious option. It offers a glossy, sleek, and seamless beauty that is incredibly functional because of its durability. A cost-effective option is available in the form using granite tiles. The same durability and beauty is available, and when installed, the seams are hardly noticeable.

Soapstone

This material, which is quarried like marble and granite, is rising in popularity. Early homebuilders crafted do-it-all sinks out of soapstone. Today, it is a popular choice due to its durability and longevity.

Limestone

This material, while extremely attractive, is not considered the best choice for counter tops. Not only is it a soft stone, but also it is extremely porous. Unlike soapstone, you cannot sand off or scrub off stains from limestone counter tops. There are some homeowners who love this kind of patina, while others find it frustrating to deal with.

Marble

This option is another tricky counter top choice because, while beautiful, it can scratch and stain easily. It is not as durable as granite and does not react well to chemical cleaning agents. These counter tops are best suited in light-use bar areas in entertainment areas of the home.

Wood

This is a functional choice in that it can be treated with sandpaper if the surface becomes scratched or stained. These counter tops are durable — particularly for those who do a lot of chopping — and easy to maintain. In comparison to other counter tops, though, they do require more care and do not last as long.

Stainless Steel

If you are looking for a professional restaurant-kitchen style look, then stainless steel may be the perfect choice for your counter top. Stainless steel is an especially clean choice because it is not porous, therefore bacteria does not grow as rapidly. It also does not chip or scratch and requires little maintenance. There are a few cons to be aware of, though, before making your final decision. If the counter top is not attached to a strong base, it can be loud from resonating water, pots, pans, and vibrate with an eerie buzz when the spray head is used. It also is not resistant to scratching and, for those with young children, it might not be the best option because fingerprints show up.

Laminate

These counter tops are among the most popular choices because of their affordability and their ability to mimic most popular stone patterns. In addition to these pros, you can also install this type of counter top yourself. Laminate counter tops are both attractive and durable. Be aware, though, that darker color choices will reveal scratches.

Ceramic Tile

For those are interested in a vintage look, creating a counter top from ceramic tile might be your best bet. While the larger tiles are not as durable, mosaics are less popular due to their complexity and maintenance issues. It is rather difficult keeping the grout lines clean, but the surface can handle a lot of wear. The tiles scratch and chip easily, however, so they are not optimal for long-term use.

The choices in kitchen cabinetry sport an equally impressive list. There are various types available, so take your time making the right choice. We will look at the different options available to help make the decision process a bit easier. This is another area of the construction process where space planning is a must. Here are some considerations you will need to make:

- What kind of wood do you want the cabinetry to be constructed from?

- Do you want a specific kind or style of hinges, knobs, and drawer pulls?

- Are the cabinet doors going to be framed or paneled?

- How large do you want your sink to be?

- Will you need an additional sink in an island?

- Do accommodations need to be made for dishwashers, trash compactors, bar refrigerators, and other appliances?

How your kitchen looks when finished rests on the quality of the materials chosen. Selecting the best materials, the best designs, and the best functionality will increase the longevity of these investments. Be aware that just because initial costs are lowered when choosing lower-quality materials, it may cost more in the long run to replace them sooner.

Money-Saving Tip:
Often, owner-builders have been selecting Corian™ counter tops (a trademark of DuPont™) because they are affordable, have been around for about 40 years, and are difficult to ruin. Corian™ counter tops provide an extremely sturdy and durable surface and are available in a large array of color choices.

11.3: Bathroom Vanities

Once the flooring is installed in the bathroom, the vanities can be installed. To save money, use the same contractor for both your kitchen and bathroom installations. Or, with some time and patience, you can complete this project on your own. For this project, you will need the following:

- Drywall screws
- Slip joint pliers
- Tape measure
- Carpenter's level
- Stud finder
- Carpenter's level
- Stud finder
- Shims
- Utility knife
- Hole saw
- Power drill
- Caulking gun and caulk

Notes From the Field

Some owner-builders have reported how gratifying it was to take hands-on classes and workshops about tile installation. It allowed some owner-builders to complete tile installation projects on their own and save money. In other cases, owner-builders noted how it was an eye-opening experience that drove them to hire a contractor.

The complexity of this project depends on the type of vanity you plan to install. The directions that follow are for a basic vanity installation. If you feel you are unable to complete this project on your own, enlist the help of a handy friend or family member or hire a contractor. Follow these guidelines for installation:

1. Measure the height and width of the vanity. Then, mark where you would like the vanity to be placed with vertical lines.

2. Measure where all the plumbing comes through the wall, then mark it on the back of the vanity (if there is one). Use the hole saw and cut out holes for each area.

3. Verify where at least one stud is in the wall and mark it off. Drill a pilot hole for the 3-inch screw that will hold it in place.

4. Remove all doors and drawers from the vanity before installation to reduce chances for damage.

5. Be sure the cabinet is level from front to back and side to side. Use shims to level it off if necessary. Once the cabinet is level, attach it to the wall using 3-inch drywall screws.

Owner-builders are notorious for leaving out bathroom cabinets during the planning process. Bathroom storage is a high priority, particularly to owner-builders with growing families, and it is equally as important to have easy access to everything. Here are some considerations to keep in mind when making the buying decision:

- **Cabinets:** They are not necessarily just for under the bathroom sink vanity. They are also available for over-the-toilet storage, above mirrors, in medicine cabinets, and in recessed cabinets.

- **Open shelving:** These offer decorative elements to the room and are often an after-thought. They can be added above bathtubs, over toilets, above (or below) mirrors, and next to sinks for added storage.

- **Linen closets:** This is a must-have in any bathroom. Choices range from freestanding closets to recessed closets.

Money-Saving Tip:

Save money by choosing an unfinished vanity. Then, either stain it or paint it yourself. This is also a good opportunity to bring custom colors and finishes into the bathroom.

CASE STUDY: OMNIARTE DESIGN

www.OmniArteDesign.com

erinn@omniartedesign.com

Erinn Valencich -- Interior Designer and TV Host

Jamie Fisher -- Assistant and Photographer

"We had two months to do the bathroom and the kitchen. It is an aggressive timeline, but very doable if you have all your materials picked out and ordered in advance. The framing period is the most interesting — at one point you could see completely through one side of the house, from the kitchen through both baths. I love choosing finishes, but it is certainly the most nerve-wracking. Especially tile — it is the most permanent thing you can put in your home. However, I think it deserves to be something you absolutely love — then you can not go wrong. Spend more than you normally would, as this is something you are going to look at daily for years to come. We chose a unique shaped tile from Ann Sacks for the flooring, and that was where we started with each room. I think it is best to start from the ground up."

11.4: Finish Plumbing and Electrical

The last plumbing phase involves all the final hookups for appliances with a water supply or chain, including sinks, faucets, drains, toilets, the dishwasher, bathtubs, and showers, among others. This work cannot occur until all the other elements of the bathroom construction are complete.

Under normal circumstances, code enforcement agencies and authorities require specifically that licensed plumbers must perform the work, and all the work must pass a final inspection. However, there are some states that allow the owner-builder to complete all of their piping, plumbing, and hook-ups. It is imperative that this information be verified depending on where you live and what regulations must be followed.

Remember, when plumbing does not work properly, everyone suffers. Not only can improper installation be costly, it can also be a huge inconvenience. When plumbing is working efficiently, it is often taken for granted. That is no excuse, though, to overlook crucial details. Prepare an owner-builder inspection checklist to keep yourself on track and so you will be aware of what your plumber should be doing. This checklist should contain:

1. Plumbing item specifications
2. Plumbing drain specifications
3. Pipe insulation
4. Miscellaneous items specific to your home

In addition, you will want to make sure the following items are addressed. A licensed plumber should abide by these accepted principles:

1. Are there enough pipe hangers in use?
2. Do the pipes run parallel to the wall studs?
3. Are there enough shut-off valves installed?

4. Did anything leak when it was inspected?

5. Have the outside faucets been installed and tested?

6. Do all the toilets flush properly?

7. Are any of the appliances chipped or damaged from installation or testing?

8. Does the water heater run efficiently?

The final electrical phase involves the installation of all breakers in the electrical panel. Like with plumbing, electrical is often taken for granted unless something goes wrong. This is an area that needs specific attention and, again, normally requires a licensed professional.

Notes From the Field

Do not expect this process to go smoothly. Contractors will be in each other's way and constantly bump into each other. Expect one or two of them to get frustrated and say they have to return at a later date. As annoying as this is for owner-builders on a tight schedule, it is common. Knowing this ahead of time, though, should allow for extra padding to be added to your timeline.

Because electrical work is complicated (and better left to a professional for various inspection, safety, and homeowner's insurance reasons), these are some basic guidelines to consider when the electrical work is being completed:

1. Spend the money on good- or high-quality wiring.

2. Make sure tree limbs and other plant growth are cut away from ground wires to prevent obstruction.

3. Do not forget to request electrical outlets on the exterior of the house, as well.

4. Consider the installation of whole-house surge protectors. These can protect your appliances, entertainment equipment, and alarm system, in addition to your computer and peripherals.

5. Be sure there is enough outdoor lighting.

You will be performing the final inspection of your home before the certified building inspectors come in. Your checklists need to be as specific as theirs are to ensure quality work and to help speed the process along when they arrive for their appointment. Many municipalities will provide a copy of their inspection checklist to homeowners prior to their visit. That way, everyone can be prepared for every detail of the process.

Money-Saving Tip:
Rather than waiting until the job is finished, conduct mini-inspections several times throughout the process. This will prevent the costs associated with labor to fix any issues that may arise due to miscommunication or the wrong materials shipments coming in unnoticed.

11.5: Stairways

There are some general standards when it comes to stair building, including angles, treads, risers, widths, landings, and framing. Prior to adding a permanent staircase leading to a second floor or down into a basement, some owner-builders choose to construct a temporary staircase. This allows for high traffic with heavy boots without ruining the stair treads. Keep in mind the rough work will be ugly, but it will serve its function until the permanent structure is ready.

When building a temporary stairway, be aware there are still requirements. For temporary staircases, build a 30-inch landing in the direction where there is the most foot traffic; the stair treads and risers in this are also must be uniform (there is a one-quarter-inch allowance in variations). These requirements are meant for the safety of you and the workers.

Notes From the Field

If you are unsure about the measurements or anything else to do with building the stairways of your home, consult your lumber supplier. In some cases, they sell pre-cut treads and risers that are ready to go. In other cases, they may custom-cut the material to order. Inquire about these options.

When you are ready to build the permanent structure, the first step is determining what kind of stairway you like. There are several different types to choose from, but not all styles will work in all floor plans or house designs. This is another example of where advanced planning pays off:

• Straight
• Spiral
• L-shaped
• U-shaped
• Circular

The next step is determining the measurements for the stair treads and risers. The treads are the length of the stairs, and the risers are the height of each step. Stringers hold the weight of the stairs and are where the risers and treads are attached. In standard construction, stair treads are normally ten inches; however, the risers are a different story. To determine the height of the risers, you have to look at how much room there is from the top of the staircase to the bottom. There is a formula for figuring this out, so if you are not good at math this will be a good time to call in some help from a friend or a family member to give you a hand. The formula for figuring this is to divide the total distance from the top floor to the

208 The Complete Guide to Building Your Own Home

bottom floor by seven. This will tell you how many risers will be needed. Once you figure out how many risers will be needed, divide that number by the height of the staircase. This will tell you how many inches high each riser should be. For example, here is how the equation would look with hypothetical numbers:

Height of entire rise, including thickness of floor: 96 inches

Divide the height by seven: 96 / 7 = 13.71 (round it up to 14)

Divide the riser number by the height: 96 / 14 = 6.85 (convert to inches)

This formula gives you the exact measurements for each of the risers. Remember, the treads need to be around ten inches. Another point to remember is you will have one less the amount of treads compared to risers, so you need to find the total length, or run, of the stairs. To find this number, multiple ten by the number of risers minus one; or in the case of the example above, by 13 risers. This tells you the run will be 130 inches. Use a carpenter's square to mark the ten-inch tread and the amount of inches for the riser, and it should come out perfectly each time.

For easier construction, most carpenters add the risers first, then install all the stair treads. This will allow the tread plate to lay flush against the riser without any gaps. Each plate must be screwed in place because, in this case, nails will not hold as well. If the treads do not hold well, they can start to separate and cause a tripping hazard.

Do you need a landing? These are normally used when space is limited. Even in homes where space is not an issue, landings are used to add interesting architectural features and make the staircase less intimidating. Another mathematical formula is necessary in order to successfully add a landing into your stairway design. Here is an example of what you will need to do, again, using hypothetical numbers:

7.5 x 3 = 22.5 or 7.5 x 4 = 30

These numbers are based on the height of the risers (7.5 inches) multiplied by how many stairs you want to climb before reaching the landing. The landing must be a square construction. If you do this measurement, the landing will fit in perfectly each time.

It is important that the staircase is wide enough for two people to pass each other. Equally important is the ability for furniture to be moved up and down the staircase with minimal trouble — if, of course, space permits. When choosing materials, think about whether you want them covered with carpeting or if you prefer the wood be left exposed. If you choose to leave the wood exposed, be sure to select a hardwood. This will increase the staircase's durability and add to its aesthetic, so it is beneficial to choose a quality wood.

The next thing to think about is the railing. Should it be ornamental? Should it be made only of wood? Before considering these questions, the first priority should be how sturdy the railing is or needs to be. Railings normally are installed when the rest of the finish work for the house is being completed. This will prevent it from being damaged while the rest of the house is being constructed. In the meantime, brainstorm options and considerations in your dream notebook.

A newel post will be necessary, either attached to the floor joists or to the stringer. The newel post should be made from solid wood and secured with lag bolts. You may wish to cover this post over with something to protect it from damage during the rest of construction. Newel posts are often secured to floor joists with bolts. Ornamental and architectural features can be added during the finish work process.

Following this, the handrail is secured to the newel post with lag bolts. You can avoid splitting the wood by drilling pilot holes first and then attaching the handrail with the lag bolts. Do not forget to use a countersink bit so the head of the screw is hidden once it is installed. If you are installing balusters, the handrail will cover them up and the screw holes. Use carpenter's glue for a secure fit.

Exterior stairways are normally not a difficult project, particularly for the experienced do-it-yourself-er. To determine the width and the slope of your staircase, you must adhere to local building codes. You can obtain this information online or from your local code enforcement agency. Regulations for how the stairs are braced and assembled are also strictly enforced, so it is vital to get all this information beforehand. The same math ratios can be used for building exterior stairs as with interior stairs. The biggest difference is going to be the design of the stairway.

Money-Saving Tip:

Be sure to spend some time price-shopping for the materials on your list. You will find that it may be necessary to visit more than one supplier in order to get the best prices. Do your homework.

CHAPTER 12
Lighting, Hardware, and the Accessories

12.1: Types of Light Sources

The right lighting choices can add just the right touch to any home's décor and ambiance. To create a room with an inviting atmosphere, it is important to be sure there is general lighting, task lighting, and accent lighting. These lighting situations can be achieved through a variety of lighting sources. Do not forget to embrace natural lighting options, as well. This book explores five sources of light.

Notes From the Field

Owner-builders looking to continue with building projects after the build have found it cost-effective to turn skylights into doghouse dormers[1]. They did not have room in their budget for the dormers in their initial build, but they would have new funds in the future. Because the roof was already opened up, they saved on the cost of renovations.

---

[1]Doghouse dormers look like doghouses sticking out of slanted roofs, providing a great source of light and adds aesthetic value.

The Sun

Sunlight, or natural light, is plentiful throughout the day and does not cost anything. In addition to lighting the house, there are also health benefits derived from natural light. Scientific studies have shown significant changes in mood and the desire for physical activity as a result of a home lit well with sunlight. Positioning of the house and of the windows, as well as the number and size of the windows, all must be taken into consideration in order to take full advantage of the sun's light.

Light bulbs

Electric lighting can be achieved through use of two different kinds of bulbs: incandescent or fluorescent. Manufacturers offer a wide variety of lamps, fixtures, and lighting controls to create task lighting, general lighting, and ambience in the home. These are the most popular consumer choice.

Compact fluorescent lighting (CFL)

This lighting source is costly upfront, but can be approximately four times more energy efficient than comparable incandescent bulbs. CFLs have many benefits over incandescent light bulbs. Not only do they produce more light per watt, they also last up to ten times longer, produce less heat, and use up to 75 percent less energy. The down-side is that they take some time to reach peak brightness.

Skylights

Not only do they provide natural sunlight, but they also offer energy benefits though use of solar heating. Positioning skylights in cathedral ceilings, in rooms with poor direct sunlight, and in master bathrooms are all options to address in your home's design plan.

Tubes and tunnels

Sunlight tubes and tunnels, also called tubular skylights, are another natural light source that allows light to reflect through a cylindrical tube. They are easier to install than skylights and resemble conventional lighting fixtures. Some tubular skylights can be regulated to change how much light is allowed to reflect through. Others have lights integrated within them so they can be used both during the day and at night. These light sources are great additions to small bathrooms that have no wall space for a traditional window.

Money-Saving Tip:

Plan your lighting well because not every room will need the same. A good lighting plan can reduce costs, but also provide all the light necessary from room to room. In addition, consider adding junction boxes that are wired and sealed with a metal faceplate. These will serve you well for future light fixture installations that you either do not have time for during initial construction of the home, or if you do not have room in your budget. Taking this step will also save you electrician fees later on.

12.2: Home Lighting Uses

It is impossible to assume homes will be using only one type of lighting system, or one type of light bulb for that matter. Lighting choices will mix both incandescent and fluorescent lighting choices. The purposes of these lighting fixtures will range greatly from room to room in the home. Be aware of these facts when setting up your lighting plans.

Notes From the Field

One owner-builder did not have enough room in his budget to perform a lot of fixture installations. He made up for that by adding less expensive electrical outlets. Then, as funds were available, additional lights were added to the rooms.

Here are some uses you will need to address:

General and Background Lighting

General or background lighting is normally created for general living activities without being dominant. Size the decorative space to ensure the proper size fixture is used. General lighting tends to illuminate an entire area, creating a comfortably lit area to walk around safely. This light use is one of the most fundamental in a home. Remember, there will be situations in which more than one fixture is needed to achieve general lighting goals.

Local and Task Lighting

Local or task lighting is reserved for specific work or activity areas. This type of lighting provides added energy efficiency and a more comfortable work area. Generally, task lighting is used where reading or detailed work is being completed as in an office. Task lighting is also found in specific kitchen areas. If you do not account for local or task lighting sources in your lighting plan, it will quickly become apparent how much of a necessity they are when you start to utilize these areas.

Accent and Decorative Lighting

Accent or decorative lighting is not meant to serve any other function aside from creating a mood or atmosphere. Commonly, accent lights feature a piece of art or an architectural element in the room. This type of illumination is frequently three times as bright as ambient lighting. You can find accent lighting on tabletops, in bedrooms, hallways, living rooms, and bathrooms. As you can see, accent lighting can be used pretty much anywhere in the house.

Dimmer Switch Lighting

Dimmer lights are a cost-effective way to prevent maximum light use. Dimmer switches allow the lighting to serve multiple functions, such as task lighting and general lighting. The allowance for lamp dimming extends the life of the bulb and also saves on energy costs. Dimming lights can create an interesting mood in a room while extending the life of the light bulbs.

Under-Cabinet Lighting

Under-cabinet lighting is a cost-effective way of creating task lighting that does not illuminate the entire room. It is inexpensive to install and saves money because they utilize less energy. Kits are readily available in most home improvement stores. Under-cabinet lighting adds functionality to small kitchen spaces by illuminating poorly lit areas. Do not limit this lighting option to just the kitchen, though. Wherever there are cabinets, these lights will work to illuminate the surfaces beneath.

Automatic Lighting

Automatic light switches, also known as motion-detecting lights, are common outdoor installations for those who are seeking the added security through use of a motion sensor, and the added cost benefits of the outdoor lights only being on periodically. These lights can also be used as a decorative feature around gardens to feature large plants or trees and to illuminate features of a home. A second automatic light fixture is one that can be set to turn on and off according to a timer, or one that comes on at dusk and off at day break.

Deigning a lighting plan can be tricky, primarily because a lot of advanced thought must be accomplished. The need to think ahead about what you want in your home, what functions a particular area of the home will serve, and where decorative elements will be are just some of the design considerations you will make. Because a lot of hard wiring and permanent installations will be made, it is important to iron out every detail well in advance.

Money-Saving Tip:

Allow for natural light flow in your plans. Consider skylights and solar tubes for maximum use of natural light. These designs will save on energy bills. You may also be able to benefit from the solar heating benefits natural lighting offers, thus saving more on your bills.

Lighting is more than a way to see in the dark. Fixtures can provide an ambiance and set the mood. Subtle lighting, focus light, colors, and brightness are as important as the furniture. Whether you hang an antique altar piece as a chandelier in your dining room or a mock Tiffany lamp is not as important as what it emphasizes. Each room needs to be distinctive and reflect your personality, as well as the brightness it brings to the room.

Probably the most overlooked expression of personality is the lighting found in bathrooms. People often spend a great deal for pedestal sinks, whirlpool tubs, and steam showers and then skimp on the lighting to see it all. However, even if you do not go in for a lot of extras in the bath area, spending a little time and a few extra dollars on fixtures and lighting can turn a run-of-the-mill bathroom into a showcase.

When the time line starts to shorten there are ways around rushing a task. Erinn Valencich shared some of her thoughts on bathroom finishing:

CASE STUDY: ERINN VALENCICH

Interior Designer and TV Host
OmniArte Design
www.OmniArteDesign.com

"Lighting in the bathroom is incredibly important. I opted to not hang a pendant light, even though that is one of my favorite things to do in a bathroom. Instead, I went with one single Costrutto sconce from Sonneman, mounted on the mirror. I loved the open square shape and felt that having just one made it stand out even more. For the hardware, I turned to Atlas Homewares. They make some of the most fun handles and knobs on the market, and they always arrive very quickly. Like most people, when you find something you like, you do not want to wait six to eight weeks, especially for a small detail like hardware."

12.3: Lighting Terms

The following is a breakdown of the various lighting terms you will use and encounter when setting up the lighting plan for your new home. This list will give you a better understanding and idea of what you will need and where to position it in your new home. This list will also be helpful when you are speaking to design professionals and electricians about your lighting plan.

Notes From the Field

Owner-builders have reported that once they had a firm understanding of the lighting terms, they were able to make a better buying decision. In some cases, it solved confusion by preventing the wrong kind of light fixture from being purchased.

Here are various lighting terms you should be familiar with:

Semi-direct lighting: This kind of light is positioned (normally by 60 percent) downward. This type of lighting is recommended when an entire space needs to be lit evenly.

Semi-indirect lighting: This is lighting that is positioned (normally by 60 percent) upward. This type of lighting is recommended when a particular zone of a room needs to be illuminated.

Valance lighting: These lights are normally a shielded panel, or other type of decorative molding, installed along the top of the window. Light is diffused upward and downward simultaneously. This is considered an indirect lighting option, and it is appropriate in just about any room of your new home.

Indirect lighting: Nearly all the light is positioned toward the ceiling when indirect lighting is introduced to the room. The light diffusion allows for no glare.

Direct lighting: When direct lighting is introduced to the room, nearly all the light is positioned toward the floor. This lighting is reserved for detail-oriented tasks.

Accent lighting: This is the lighting created to illuminate a particular object or area. Accent lighting is frequently found on tabletops in bedrooms, bathrooms, hallways, home offices, and just about any other room in the house. Normally, these lights are small and serve mainly as a decorative purpose or to add ambiance to a room.

Cornice lighting: This lighting is made up of a panel attached parallel to the wall and the ceiling, distributing light downward only. This type of lighting is used to dramatize drapes, artwork, and other wall treatments. Architectural elements are also accentuated through use of cornice lighting.

Cove lighting: These lights cast illumination toward the ceiling from a shielded source. This lighting is perfect for accenting ceilings and other architectural elements of the room. Cove lighting can provide both general and accent lighting, depending on where you would like to use it in your home. This mode of lighting creates a gentle light wash over the ceiling, making for a comfortably lit space.

General diffuse lighting: This is an illumination source that casts light equally in all directions. Suspended globes are a good example of general diffuse lights.

Luminous lighting: A light source mounted into the ceiling with a diffusing shield. This type of lighting option is normally used in basements with suspended ceilings. However, they can be employed in kitchens to provide additional general lighting without use of an obtrusive ceiling fixture. Luminous lighting also works well as a faux skylight in dark closets and bathrooms.

Recessed lighting: Lighting that is placed flush against the surface of the ceiling, with only the light bulb and ring holding the fixture together showing. These lights can be used in virtually any room in the home because they blend so well. Recessed lighting generally points the light in a downward position, focusing in one particular area.

Track lighting: is a system of fixtures that can be moved and positioned along a track to meet your unique and individual needs. Track lighting can be both on a track and recessed. This is a popular choice due to its versatility. It can be used to highlight artwork, architectural elements, and other interesting elements in the room. It also serves well to illuminate the room with general lighting.

Wallwashers: These lights have baffles or reflectors, allowing them to wash the entire wall with light. Normally, when a wallwasher is in use, the texture of the wall is diminished. This provides a powerful presentation in the space.

Down lights: Are lights that are hidden and deliver a spot of illumination to a specific area downward only. This lighting option can either be recessed or non-recessed, and may either be a feature or discreet. Downlights are often referred to as recessed lights because they serve many of the same functions.

Up lights: Lights that are often hidden and deliver a spot of light to a specific area upward only. These can be freestanding or an installed fixture. Normally, the majority of the light (90 percent or more) is directed upward. This type of accent lighting is normally placed behind plants, sofas, under glass shelving, and anywhere else a dramatic accent look is desired.

You are likely to come across other lighting terms that are not mentioned here, but this should get you started. If you are unsure if the lighting choices you are making are correct because you cannot find them on this list, consult with a design specialist at your local home improvement center, an architect, electrician, or the Internet for more information.

Money-Saving Tip:

It is not necessary to purchase a lighting guide or a book about lighting in order to understand what all the lighting terms mean and what they are best-suited for. All this information is available on the Internet for free, as well as in hand-outs provided by most home improvement centers and lighting manufacturers.

12.4: Hardware: Doors and Drawers

Putting the right finishing touch on your bathroom or kitchen can be as simple as the choices made in door and drawer hardware. Take this opportunity to show off your own unique style and be creative. There are significant questions to ask yourself when determining what is right for you when it comes to hardware:

1. What styles of drawers and cabinets do you like?
2. What colors of drawers and doors do you like?
3. How many drawers and doors do you need?

There is a plethora of options available in terms of finishes, styles, and colors available for door hinges, doorknobs, and drawer pulls. Do you want them all to match throughout the house? Do you want the bathrooms to be a different style than the kitchen? Do you want all the doorknobs in the house to be the same? Asking yourself these questions will save you a lot of time when it comes time to shop for these items.

There are some basics about hardware you may or may not know. A "pull" or a "handle" refers to hardware with two mounting points (two screws attaching it to the surface). A knob normally refers to a piece of hardware with one mounting point. While there is no hard and fast rule for installation, most situations allow for handles to be on drawers and knobs to be on doors. However, there are beautiful designs where pull-type handles are installed on the cabinet doors. This is a popular option in modern, contemporary, and traditional kitchen and bathroom designs.

Common mistakes can be made when choosing hardware. Tips helpful for preventing these mistakes include:

1. Do not rush the decision. Take your time finding exactly what you want for precisely the look you are trying to achieve. This tends to be the No. 1 mistake owner-builders make.

2. Do not be a follower. Just because your neighbor (and everyone else) has chosen a specific kind or style of hardware does not mean that is the right choice for your home. Do not be afraid to try something new, explore different options, and be unique.

3. Be aware of design features. The last thing you want to do is choose the wrong kind of hinge for your cabinet design. Take accurate measurements, figure out what finishes you want consistent throughout the design scheme, and pay attention to colors.

4. Watch the size of your knobs and pulls. If you do not take into consideration the size of the facing, you could choose a pull or a knob that is too small or too large. This is why accurate measurements must be taken.

5. Do not skimp. If you purchase the lowest-priced item, there is a chance you are sacrificing quality. Lowering initial costs are good in some instances, but not in all. You want longevity and durability.

6. Do not over-spend. There is a middle ground for hardware, so shop around. Top-of-the-line products can sometimes cost inflated prices, therefore practice buyer-beware principles. Focus on the materials you are shopping for, and use your best judgment.

7. Do not focus on what cannot be seen. Drawer sides, for example, are not visible, so it does not matter what they look like. Consequently, it is not necessary to have the entire drawer constructed from your favorite wood.

8. Do not spend extra for optional features. If the drawer stop is optional, do not add it to your purchase. These are things that can be added later when maintaining a home building budget is not your focus.

There are other optional pieces of hardware you will be tempted to purchase. Remember, though, they are not required and that you are trying to keep your budget in check. For example, do you really need a back plate for your drawer pull? While it is a beautiful decorative element that will surely be a crowd-pleaser, it is not necessary when it comes time for the home inspection or your loan closing. These things can be added in later — add it to your Christmas list.

Money-Saving Tip:
Expensive door hinges do not need to be professionally installed. High price points can be avoided, too, by shopping around.

12.5: Interior Priming and Painting

If you have the time and some extra help from friends and family, you can save a lot of subcontracting dollars performing the priming and painting on your own. More money can also be saved by putting off the painting for a period of time and only doing the priming to receive your final inspection. Clean all the brushes, rollers, and paint trays thoroughly when the priming is finished. That way, if you plan to paint later after the inspection, they will be ready to use when you are ready to complete this step.

Priming is done to make the paint adhere to the surface. It does not require as much care as the paint does, but it is applied in the same manner. Start with the ceiling by focusing around the edges and fixtures. Then, roll out the rest of the surface. In nearly all cases, one coat of primer is enough. Some recommend using a primer that contains a sealer for new construction. If you are unsure of what product to choose, consult your local home improvement center or your contractor for advice.

Notes From the Field

It is not uncommon for a contractor to recommend that homeowners wait at least one year following the completion of their home before painting any of the rooms. During the first year following construction, the house is going to settle, causing distress cracks in the drywall. It is a lot easier to repair these cracks and then prime back over them than it is to deal with sanding, repairing, and touching up a painted surface.

After the ceiling is finished, move on to brushing the edges of the walls and around the fixtures. From there, roll out the flat surfaces using a "W" formation when applying the roller to the wall. Once the primer has completely dried, tape off the ceiling and any other areas where you do not want paint color.

When the room is masked off, cut around the edges and the fixtures. Do those close areas with a brush, and then move on to the roller. Use the same "W" formation when rolling the paint on to the surface. Depending on your color choice, it might be necessary to apply more than one coat. Do not rush through the painting — it will look messy, paint will splatter on to the flooring, and areas you want to leave unpainted could end up with paint on them. Take your time and always use a drop cloth and hat.

Here are some basic rolling tips:

1. After adding the paint to the tray, roll the roller back and forth from the paint to the ridged area of the paint tray. This will remove excess paint from the roller and prevent splattering. If you feel there is still too much paint hanging on the roller, use a paintbrush to wipe off the excess.

2. Start painting in a zigzag or "W" formation first. This allows for even coverage and prevents streaking from occurring.

3. Follow this pattern with long, up-and-down paint rolls. This will finish off the coverage, as well as even out the color.

4. Paint small sections of the wall's surface at a time. For example, try to stay between 3- and 4-foot areas. If you are working with a particularly large wall surface, this could become cumbersome. So choose a section size you are comfortable with.

5. Paint the walls from top to bottom for the best results.

Select your paint and primer wisely. Inexpensive products may be less costly by the gallon, but you will need more of them. You may need to double the cost of the paint and primer and the number of coats. High-quality paints (including some store brands) generally have two advantages: they paint more per square foot than the inexpensive brands and usually require one coat for coverage. Also to be considered is that they stand up to washing — something mandatory with children and sticky fingers, crayons, and toys bumping the walls.

12.6: Appliances

The refrigerator: Refrigerator choices are plentiful, so it is beneficial to weigh all the options available. Not only do space constraints and energy efficiency play a role in the decision-making process, but the style of the refrigerator is also a crucial element. Beyond the typical standards, there are a number of finishes to choose from, ranging from stainless steel to refrigerator trim kits. Here are some options to consider:

- **Top-mount freezer refrigerator:** The top position of the freezer allows for easier access to frozen food. These models are most common in model homes and those seeking cost-effective alternatives during the initial build.

- **Side-by-side refrigerator:** Offers more storage space and has a reputation for being reliable, convenient, and attractive. These models have also been reported to save on energy bills in comparison to top-mount freezer refrigerators.

- **Freezer on the bottom:** This is more energy-efficient than any other model, and because the food is at eye level, food is easier to access. Such models are growing in popularity rapidly for those interested in energy and financial conservation.

- **French door:** The primary features of this model are a side-by-side opening refrigerator on top and a freezer on the bottom. This relatively recent innovation has been growing in popularity for its functionality and efficiency.

The Oven: Even if you are not a top chef or a recipe connoisseur, oven choices are still significant ones in the design and functionality of your kitchen. The style decision is going to weigh heavily here. Choose what will fit your lifestyle, space constraints, and budget the best. Oven options are also numerous:

- **Single oven:** This is the most commonly styled oven found in homes.

- **Double oven:** allows for two different foods to be cooking at the same time at different temperatures.

- **Convection:** Hot air is circulated with a fan. The fan intensifies the heat as it circulates allowing the food to cook faster and more evenly.

- **Conventional:** Another common choice found in ranges. These ovens do not contain a fan; therefore, heat is not evenly distributed throughout the oven.

Cook tops: The type of cook top you choose will be affected by what kind of burner you will be operating (gas or electric). Once you know this, your choices are automatically narrowed. Select a cook top that will accommodate the amount of cooking that normally goes on in your kitchen. If you do not cook often, do not waste your money on top-of-the-line, professional cook tops. Here are some choices to consider:

- **Standard coil-burner cook top:** An economical choice, the elements heat quickly, allowing you to cook right away.

- **Ceramic glass cook top:** This style of cook top is stylish, sleek, and easy to clean.

- **Gas cook top:** If you are looking for precise temperature control when cooking, this is the right choice for you. In addition, it is easy to clean and some offer pilotless features.

- **Downdraft design cook top:** The vent in the middle of the cook top prevents the need for an obtrusive vent hood above the range.

Ventilation: Cook top ventilation offers two options for homeowners: either the updraft or the downdraft. Local building codes will dictate your choice to a point, but the kitchen layout and personal preferences will play a larger role. You will pay more for ventilation with features such as higher airflow and noise reduction.

Notes From the Field

It is possible to have a custom vent hood designed, built, and installed into your kitchen design. Not only does this increase the beauty of your kitchen, but it also boosts the value of your home. If you do not have room in your budget for this during the initial build, these kinds of additions can be made later.

Updraft vents are located in the vent hood over the range or cook top. These vents are not only extremely effective, but they also come in a wide variety of styles and models. The ability to spice up your kitchen with a decorative vent hood is not an impossible task. This is another area where it could cost a considerable amount if too many design elements are introduced.

Downdraft vents are less noticeable, but they are, on the other hand, less effective. Air is drawn downward and out through vents on the side and back of the cook top. They are less effective because they only catch polluted air directly next to the vent.

The microwave: Even top chefs and kitchen wizards need to warm up a cup of coffee or eat convenience quick meals every now and then. If you are a heavy microwave oven user, you choices will automatically be narrowed down based on your need. Microwave ovens come in five basic types:

1. **Counter tops:** The size of most of these models allow for placement in the kitchen where it is needed.

2. **Built-ins:** This microwave caters to kitchens in need of space saving appliances because they are built right into the wall or the cabinetry.

3. **Under the cabinet:** another space-saving option.

4. **Over the range:** these commonly come equipped with a range fan and lights similar to what is found in a hood.

5. **Drawer microwave ovens:** This is the newest innovation in space-saving models, designed for ease of use and efficiency.

The Dishwasher: Choosing the right dishwasher is a longer process than some might imagine because there is a lot to choose from, so have fun with this shopping excursion. Opt for a model that is quiet, features a touch pad (rather than dials), and has a sleek exterior (such as stainless steel). Dishwashers come with a lot of more expensive features now, so be sure to find something within your budget. Do not be talked into a model that is more than what you truly need for the home. Also take into account the energy used when choosing which model is right for you. This information is featured right on the machine.

The garbage disposal: Disposals come in two different styles: continuous feed and batch feed. The more common of the two is the continuous feed because it allows for drop-in use while it continues to run. When selecting the right garbage disposal, you should consider the horsepower, if it can auto-reverse (to clear jams), the noise level, corrosive-resistant shield, and the included type of warranty. Remember to take into account how large your family is and how many meals are prepared. By default, a larger family will require a larger disposal unit. Standard models come with 1/3 horsepower, but larger families should look for larger motors for longer product life.

Money-Saving Tip:

Whenever possible, look for the Energy Star logo. Purchasing Energy Star appliances can save you hundreds of dollars in energy use of the lifetime of the appliance. Many stores offer free literature with cost comparisons among various appliances when they are in use to give consumers the ability to make an informed and educated decision. Several Web sites also provide this information. Consider adding it to your dream notebook or along with any other information worksheets you have filed.

It does not matter whether you are restoring an old house or building a new one — the secret to your success lies in your creative ability. Before looking for extraordinary solutions to problems encountered, seek out the most logical and least complicated.

CHAPTER 13
The Final Inspection

If everything thus far has gone according to plan, this process ought to be relatively pain-free. Should problems arise, though, it is going to be your responsibility to find, report, and make sure the steps are taken to correct any imperfections or errors. The first thing to do is take out all the checklists, worksheets, and other record-keeping documents that have been used throughout the build. These will act as your guide for preparing for the final inspections. There will be more than one inspector involved with this process, and often these inspections occur as soon as the tasks are complete. Inspections you can expect are:

- Electrical
- Heating
- Plumbing
- Building

During the inspection process, structural concerns are placed directly under the microscope. These issues include, but are not limited to, problems with the roof construction, floor joists, and the foundation. Even when hiring builders with the best of intentions, these problem areas are not uncommon. It is not that these builders want to intentionally dupe the homeowner, but rather these are areas where mistakes tend to be made. These are areas where codes are strict and numerous, so it is crucial to stay on top of your builder and ensure everything is done by the book.

Notes From the Field

A time-saver to consider is to ask the code enforcement agent what are the most common mistakes. That way, you and the contractors can address them ahead of time to the best of your ability.

Additional areas of concern have included electrical issues. Be sure all your smoke detectors are fully functional — not only for the safety of your home, but also to pass inspection. In addition, it is important all the circuitry in the home is in full working order and all the wires are properly grounded. For this reason, many owner-builders choose to forgo the option of installing the electrical portion of the build and hire licensed electricians.

The list does not stop here. Frequently, you will find the plumbing issues that normally come up during the inspection process. Be sure water pressure is strong enough; there are no drainage obstructions in the pipes; nothing backs up other drains; and the hot and cold water is not reversed. They are not the only issues inspectors come across, though. Missing drainpipe cleanouts and vent stacks not being properly vented through the roof are other issues. Like electrical work, the details can be overwhelming to owner-builders. It is common to hire a licensed plumber rather than deal with the issues.

Heating and air conditioning is addressed next on the list of inspections gone through with a fine-toothed comb. Construction debris can find its way into vents, so be sure those are clear of any obstructions. Leaks and lack of proper support for the ductwork are other areas of concern. Problems like these can be easily avoided and, if they still occur, they are quick fixes.

Finally, the exterior of the home will be inspected. If an inspector finds siding that is two different colors, improper caulking, or water pooling in the ground, he or she will address these concerns and require fixing them prior to receiving a passing inspection. With improper ventilation or no well protection, you will have a delay. In addition to this, approval will be denied if there are loose fascia, gaps, and improper grading.

The code enforcement agent will not only be grateful for you seeing to all of these details, but it also speeds up the inspection process. They will appreciate your attention to the facets of the inspection process, as well as your respect for their tight schedule. No, you are not trying to impress the code enforcement agent with your skills and organizational abilities; rather, you are trying to make this process go as smoothly as possible.

Do not let this phase of your project frustrate you. These inspections occur not to hinder you from finishing things off, but rather to ensure the safety and quality of your home. The last thing any new homeowner wants to experience is a leaking roof or windows that allow rainwater to flow into the interior of their newly built home. So, go through the process from beginning to end and address all the concerns as they come up. Like every other challenging part of this build, this too shall pass.

Money-Saving Tip:

If you perform frequent inspections as the build is occurring, you will save time and money in the end. This will help speed up the final inspection process, as well as save you money in labor for contracting out what needs to be fixed.

CHAPTER 14

Landscaping, Your Drive, and Walkways

14.1: To Pave or Not to Pave

Budget constraints will probably determine whether you pave your driveway after the house is complete. Eventually it is an extra that will enhance the aesthetics of the building and the value of the property. In addition, it may be required by local code.

Aside from the physical appearance, driveways and walkways are a convenience in avoiding mud and other elements from the outdoors being brought inside. A finished driveway prevents damage to vehicles, and proper grading avoids flooding. You can avoid liability issues when guests come to visit so they do not trip on an uneven surface, and the ground is less likely to shift under the stress of heavy rains. For those in colder climates, snow shoveling and plowing is considerably easier with a finished driveway and walkway than an earthen or gravel one.

There are few people who are content with having earthen driveways and walks. After the first rain, most owners will tire rapidly of mud, pot holes, and uneven surfaces. There are alternatives, each with advantages and disadvantages.

14.2: What to Know About Driveways

Driveways are a two-step process. First the gravel base is laid, 6 to 10 inches, then allowed to settle for at least eight to 12 months. The gravel should extend to about 10 to 12 inches beyond the width in order to accommodate for rain and ice. Failure to abide by these guidelines may give rise to minor cracks in cement and asphalt applied over the crushed stone.

Design is limited by your imagination, space and, to some extent, the material you use. As a basic rule of thumb, there are minimum widths for driveway length and at the curb. Single width calls for 11 feet across and 17 feet at curbside. Double width is 22 feet and 28 feet, respectively. The wider width at the street or road permits easier entry and exit, especially into heavily traveled roads. If you are going to be living in a community with covenants, be sure to see you follow their criteria for uniformity.

If you have the option of a circular drive, it may well be worth the time and expense. It affords easier access for guests, more room to park when entertaining, and reduces the chance of backing up onto things unintentionally. There are disadvantages as expense goes, but you will have to weigh the decision. A turn-around is handy, as it avoids having to back out of the drive. Try not to accept a driveway plan that only permits backing out onto a roadway, especially if it is on a blind curve or hill.

All driveways should have a pitch that takes water away from the foundation. If, for example, your garage is atop a slope leading to the street, it should still have a pitch so heavy rains do not flow down in torrents. You can always install French drainage pipes beneath the surface on the side of the grade. Such pipes are made of PVC with holes in them and are generally connected to the storm drain. The usual grade is two to three inches every ten linear feet. Conversely, driveways should never slope down into a garage. To do so would invite debris, ice, snow, and water to accumulation. If necessary, build the garage up higher.

The driveway should avoid being a straight one on a steep hill. It is dangerous to park vehicles on such inclines, and this increases the chances of someone slipping and falling on the wet or icy ground.

The finishing material, such as asphalt or cement, should be at least 5 inches thick, with the bottom edge slightly below grade.

14.3: Paving Materials

Gravel

As a finishing material, gravel should be a last resort. It is the least expensive but needs to be refreshed with full or partial truckloads of stones. The gravel often requires regrading as it shifts, as stones get crushed under the weight of vehicles, feet, and sink into the wet ground. Add to that weeds rising between the openings, and you will be busy maintaining it. Another disadvantage to gravel is that when it gets wet, it is diminished, and spinning tires can cause damage to the vehicle, as well as any object or person nearby.

Asphalt

Depending on your climate, asphalt may be a viable alternative. It is less costly than concrete and stands up well to cold. It presents somewhat of a problem in warm weather and climates due to its tendency to soften with heat. Asphalt needs to be resealed periodically or face deterioration from the elements. If you leave something heavy on the surface, such as concrete blocks or a camper, expect to see depressions. It is not recommended near pools or areas where people may walk barefoot.

The color of asphalt accents most homes well except they do not aesthetically complement concrete sidewalks. The material does not apply well to free-form designs. The process for laying asphalt by the do-it-yourselfer is quite laborious and involved heating the product, rolling it, and ultimately sealing it. It is a job probably best left to the professionals. As always, check references and never use a contractor who comes to your home unsolicited, no matter what they offer. Remember: You get what you pay for.

Asphalt is applied in two layers above the base. About 2 2/3 or 3 1/2 inches go down first, and the rest in the second layer.

Concrete

When it comes to strength, durability, and being trouble-free, concrete is the first choice for driveways and walkways. It is more costly but calls for less maintenance than other materials. Cement also pours into any design you can imagine.

Concrete is measured in terms of bags, even though you may have ordered ready-made. The mix needs to be at least 5½ bags per cubic yard of cement.

Working with concrete is quite meticulous in the set-up. In addition to the base excavation, you will also need a form around the edges. Straight lines are the easiest. You can make your own out of wood or heavy plastic. Circular or free form shapes will most likely require a flexible plastic that you can purchase at your local home improvement center. Steel rods or wire mesh is laid to strengthen the product and provide structural integrity. Cut grooves every 11 feet, using felt to line the grooves. This will allow for expansion and contraction due to climate changes and will prevent cracking. A good sealer is recommended, and also makes the surface easier to keep clean.

Notes From the Field

One caution about the care of concrete: Never use salt to melt ice on it. This can cause the chemical configuration to break down. It will result in flaking and full-thickness cracks. Use an artificial ice melter for a long life of the driveway.

Brick or Cobblestone

The most costly and labor-intensive driveway materials are also the height of finesse. Brick comes in many colors, and cobblestone is notable for its durability and old-world charm. These materials were used in all major cities until the advent of asphalt. They may be set in sand but will shift over time. The best presentation for brick and cobblestone is to be set in cement. In order to prevent fading, sealers are recommended periodically.

14.4: Walkways

Walkways allow for greater variability in material. It really depends on whether the intent is functionality, decorative, or both. Where you reside will also influence your decision. For instance, the path leading to the front door in a place like Iowa will probably call for cement. This area of the country gets heavy snow fall through the winter and then is faced with the spring thaw. Come April and May, anyone stepping on the grass can easily sink up to their ankles under the right conditions. Conversely, someone living in the high desert may do well with slate set in sand.

We have already discussed brick, cement, and gravel when addressing driveways. Yet there are other materials to consider for walkways.

Stone Slabs Such as Slate

These have the advantage of being easy to lay no matter what your experience level. The ground is prepared by digging down approximately four to six inches. A plastic liner is preferred to prevent weeds, then at least a two inch layer of sand or finely crushed stones are put on top. The slabs are set into the base and the rest of the sand or gravel is spread on top to provide a level surface by filling the spaces. It is important not to get slabs that are too thin as they will crack in handling or when weight is applied. Walkway stone should be no thinner than 3/4 of an inch unless laid in concrete.

Slate can be cut by a cement/stone blade, but to preserve the natural lines, it is preferable to score the reverse side and gently tap the area with a hammer just as you would when cutting glass.

By digging down deep enough in preparation for the stone, you can minimize the loss of sand or gravel during wind and rain. The stone may also be set in cement but as the ground shifts, you will probably see areas of separation between the edges of the stone and cement, especially if the ground is sandy.

Bark or Wood Shavings

These are a sometimes-popular choice for the first year, but then the owners realize all the problems that come with the inexpensive but good-looking choice. The primary problem is termite attraction — and it is a small jump from the walkway to the frame of your house.

In wooded areas it can become a potential fire hazard, and it also provides an uneven walk surface.

Where To Place Walkways

You will want to provide easy access to your front door, avoiding the necessity to walk on the grass or soil. Other areas to consider are those that are well-trafficked, such as from the driveway to the back door. If you have flower beds or other areas visitors might appreciate seeing and you do not want them trampled, a pathway may be in order. It can be decorative with slices of a tree trunk, large ceramic tiles found in garden shops, etc. Your imagination and budget are the only limiting factors when it comes to decorative paths.

Money-Saving Tip:

There are a number of new products on the market utilizing recycled material such as glass and tires. Studies indicate they are stronger and more durable than the old stand-bys, often costing the same or less.

14.5: Traditional or Xeriscaping

Be realistic in your expectations when it comes to landscaping. Use of indigenous plants will prove hardier than those used in other climates and altitudes, and they will generally be less expensive than plants not grown locally.

Notes From the Field

Contact your county or state extension office to discuss local vegetation and planting tips. The expert advice is free, and they can often lead you to sources of free or low-cost items such as seedlings.

Purchase grasses that require less attention, grow thicker, and are drought-resistant. They might not be the least expensive, but you will not have to reseed every year if you get a grass that works. The thicker your grass, the less of a chance that weeds will grow. It is necessary for you to provide vegetation in open areas or suitable ground cover such as crushed stones to prevent soil erosion, flooding, and water damage. This is of particular importance if the home is situated on an incline.

Xeriscaping is rising in popularity because the vegetation requires minimal water and often less attention. Traditional gardens require a significant supply of water. That may not be a problem in an area like Seattle, but certainly is a problem for those in Tucson. Aside from the cost associated with metered water, there are the ecological considerations. Xeriscaping utilizes drought-tolerant plants, low flow irrigation, and recycling water when feasible.

You can have the best of both worlds by planning. Those who are installing a sprinkler system can use low-flow heads. Plants get a constant flow of water for a designated time and are better fed than with faster, more-volume systems. Select vegetation that does not require high amounts of water and you can still have a colorful, attractive garden.

If a sprinkler system is not in your budget, plan your gardens around where the sprinklers will eventually be located; there is less interruption of your beds when they are installed. Like everything else we discussed, it is all a matter of planning. Until a system is installed, try using soaker hoses on a timer. They use less water than traditional hoses and are more efficient, allowing plants to absorb the liquid slower and deeper. There is also less likelihood of sustaining run-off from soakers than the traditional hose and sprinkler head.

Money-Saving Tip:

When you start your garden, consider planting perennial plants versus annuals. Perennials may cost slightly more, but do not have to be replanted yearly. Annuals are just that: They live one year and are replaced. You do not have to buy all your plants the first year. Leave room in the beds for additions in subsequent seasons. Some of the best deals on plants come in the fall and spring at farmers' markets. The fall is also a good time to obtain seeds and planting pods at discount prices from the superstores. Try starting seeds indoors during the winter to have your own nursery items in the spring.

14.6: Ponds and Pools

Ponds and pools do not need to be explained extensively, but will be addressed. They are more expensive and high-maintenance than you might think.

Pools

There are two common types of pools, in-ground and above-ground. These two common types have several variations. In-ground pools may be made of fiberglass, a plastic liner with a sand base, a liner with a frame and light cement base, and Gunite, which is a type of cement. All require excavation and burying the supply and drainage pipes. Above-ground pools are of different frames, and some are conducive to being taken down in the off-season.

The only difference between in-ground and above-ground is the size of pool available and cost: maintenance does not differ. Every pool requires vigilance and maintenance almost daily during the season it is used. Chemicals need to be added to maintain healthy water conditions. Temperatures can change daily, and so will the addition of how much chlorine or chlorine substitute you use. You must maintain the pH level, which is the acid-base balance, and you must maintain the algae killer. This can get expensive. Additionally, you will need to cover the pool when not in use, or suffer debris in your filters and the heat loss. The list goes on and you need to be aware of what it will take to have your pool available. Most importantly, you will need extra liability insurance. Most insurance companies, not to mention the increased number of local ordinances, require fencing safeguards to prevent accidents.

Do your homework when considering implementing a pool. You will need to know how high the water table is before deciding on an in-ground model. Know the terrain and whether there are frequent minor earthquakes. This is important to determine if the pool's structure is capable of withstanding minor shifts in soil. Some land covenants prohibit above-ground pools.

Ponds

Ponds can be even more high-maintenance than pools. Unless the pond is spring-fed, you will have to circulate the water year-round. Most people with ponds also like to have fish in them. Depending on your climate, the fish may well have to be fed daily through all months. If the pond is not excavated deep enough, the water will get cloudy. The depth should be no less than three feet at the deepest part. Ponds require cleaning and removing algae build-up. Fish thrive on algae, but too much starves the water of oxygen and the fish cannot survive. You must be careful using chemicals for the same reason.

Talk to owners of pools and ponds before making your decision. They are a great deal of fun but are a lot of work.

CHAPTER 15
The Finish Line

15.1: Preparing for Taxes

When making personal tax decisions, it is best to look to a tax professional or certified public accountant (CPA) for the best advice. Few like thinking about or paying taxes, but that does not mean this step of the process can be avoided or ignored. You must arm yourself with as much information as possible.

Property tax rates are set by your local state, county, and town assessment boards. The amounts are based on the value of your property and will rise when vacant land is improved with buildings. Generally they are rated in mils, which is one thousandth of a dollar. In addition you may pay a tax based on the amount of your mortgage, sales tax, and more, depending on the state regulations.

From the moment you purchase your land, you start paying taxes. They are considerably lower than when you finish the house, so it is beneficial to take this into consideration when setting up your budget. You will see an immediate jump in value once the house is completely finished. Speak to your town hall or city hall about what can be expected. More often than not, they can give you a basic idea and understanding of what you can expect.

Notes From the Field

Some owner-builders have to stop the construction process because they cannot afford the taxes on the finished home. Had they done their homework ahead of time, this problem could have been avoided. Use their example as a lesson to learn from rather than having to learn the hard way yourself.

To avoid the shock of what your taxes could be, develop a savings battle plan and be prepared. It is better to have too much money set aside for this bill than not enough. An option to consider is to have these taxes paid directly through your mortgage company. This way, you will never be late making a payment, and you never have to worry about forgetting to make a payment. Many lenders will insist on this method, putting money into escrow, to avoid forfeiture of their investment to unpaid taxes.

The taxes you will be expected to pay will be based on where you live. To get a good idea about what you can expect, contact your tax assessor. They may be able to give you a ballpark figure, or at least provide you with the information they are supplying to your mortgage company.

Money-Saving Tip:

It costs you nothing to consult with a professional at your lending institution about any concerns you have regarding taxes you are going to have to pay on your home and property. Many times, the mortgage company or financial institution can offer you information or tips on where you can learn more about property taxes

15.2: Punch List

The punch list is what is created before the final inspection. The list contains all the items that need fixing, addressing, or any other repairs to get the house in the agreed-upon condition of the signed contract. The homeowner creates this "inspection list" when the house is declared finished; that way, the subcontractors responsible for the particular jobs listed can address these items. Punch list items are commonly given a 30- to 60-day window of completion, which is mutually agreed upon by both the homeowner and the contractors involved during their time of hire.

Punch list items can be anything from damage to constructed items or the inability to complete final installation of building materials. Copies of this "to-do list" should be with both the homeowner and everyone else required to be completing each task. This document allows the homeowner to legally withhold final payment to the contractors until the items are completed as agreed. Be sure the copy you have is signed and dated by the contractor responsible for the fixes listed. This will serve as verification that each item in the record has been discussed, mutually agreed upon, and understood. An example of a punch list can be found in Appendix B: Check Lists.

There are numerous ways to anticipate what will go into your punch list. Like everything else you have gone through throughout the build, this will require advanced planning. How you approach this depends on your own style and method of bookkeeping. The idea is to create a record system that is easy for you to update frequently and that accommodates the space you have available to work. Here are some suggestions:

- **Create a white board or poster:** If you are working in an office with some wall space, set up either a dry erase white board or a few pieces of large poster board. In this area, attach notes and pictures pertaining to work in progress or needing to be finished, as well as any repairs needed. Neatness does not count here, as it is meant to serve as an active reminder about what to include in your punch list. Plus, the contractor may request access to this punch list area to sign off on items completed and to see the progress of others on his crew.

- **Keep a diary:** For those who prefer a relatively portable method of record keeping, writing notes and keeping pictures in a diary can serve you well. This can be anything from a daily planner with a notes section to a spiral-bound notebook. You may also opt to create a section in the notebook to bring to the job site with you daily. Choose a system that works best for you to use this tool effectively.

- **Scrapbook it:** There are those who like taking the HGTV approach to homebuilding, including creating a scrapbook of the events occurring on the job site. This is beneficial on many levels because not only does it provide a pictorial example of what is happening and what needs to be done to complete the job, it also serves as a piece of memorabilia once the job is complete. This system will require a bit more work and help, so it is not for everyone.

Above all, you must choose an option that suits your needs, habits, personality, and comfort level best, or else your ability to stick with it will be limited. As you can see, there are many alternatives available for taking an advanced approach to creating your punch list. The idea is to plan ahead, keep accurate records throughout the process, and put it all together at the end.

This punch list organizational system will save time, money, and frustrations when problems creep up later. For those facing an especially difficult time crunch, tape record what the punch list items are, as well as the conversations with contractors about them, if possible. Then transcribe this information onto your choice of a record-keeping system. Be sure to speak to the contractor ahead of time to ensure they are comfortable with being tape-recorded. Some people are not, and it is your responsibility to respect their wishes and find an alternative record-keeping method.

Money-Saving Tip:

Type your punch list out on a computer word processor and plan to print out a copy for everyone involved in the process. In addition, it will save you time when you have to add more items. Remember, time is money. For those who prefer working on paper, purchase an inexpensive notebook that will be reserved just for the punch list and label it in this manner. Each time an item is added, consult with the contractor and make them another copy, or add it to the copy you provided for them.

15.3: Moving in and Managing Your Investment

Moving into your new home takes as much planning as it does building, especially when the sale of another house is in order. If you are moving in from a rental property or another temporary housing situation, then the process is not as complicated. Let us assume, though, that the sale of the house is complete or the lease has been finalized and you are ready to move into your new home. What do you need to do to prepare? Here are some tips for planning ahead and having a smoother move-in day:

1. Fill out a change-of-address card at the post office. Ask the clerk how much lead-time they require for this change to be completed. You may also need to inquire about mail forwarding services.

2. Change the address on your driver's license, vehicle registration, and auto insurance policy. Plan ahead for these tasks because each is time consuming.

3. Transfer your utilities, phone, Internet connection, and newspaper delivery to your new address prior to the move-in date, so it is all activated at the time of the move. Some of these services require reconnection fees. Plan this into your post-construction budget plan.

4. Transfer schools, if applicable, at least two weeks prior to the move so the children will have a start date for school soon after the moving is complete.

5. If moving out of the area, find a new doctor, dentist, eye doctor, and pharmacy prior to the move to allow for having medical records transferred. Check with your current provider to see about how long this process takes, and then plan your time accordingly.

6. Put all the documents related to house construction, owner's manuals for all the appliances, and all the warranties in a storage box or filing cabinet. You will need these materials when you create your home maintenance plan.

Move-in day is going to be a chaotic frenzy of activity. Things will get dinged, dirty, and messy. Placing unrealistic expectations on yourself will only lead to stress, frustration, and disappointment. Placing these anticipations on the people who are helping you, as well, will cause hard feelings and the risk of abandonment before the move is complete. Go with the flow and be patient.

Be a smart packer. Label all your boxes with colored Xs. Then, draw out a map of your house with a colored X, marking the spot for each room. People who are helping you move in can refer to the map you have drawn and put the boxes into the appropriate rooms. When it comes time to unpack, this will help ensure there are not boxes that belong in the kitchen found in the upstairs bedroom. Prepare for unexpected things to occur, as well as for the potential for disaster. You never can be too sure what will happen on moving day.

Notes From the Field

There are many systems you could use to organize your move. Some homeowners have used colored dot stickers to designate each room and then posted a color key on the front door. The point is to find a system that you can stick with and that is easily understood by everyone involved in the moving process.

Actively delegate. Rather than hiring a moving company or any other kind of service, enlist the help of friends and family on moving day. Plan on providing drinks and meals for each of them throughout the day — it is the least you can do. Try saving money by asking friends with trucks for help rather than renting. Do not forget to reimburse them for gas money and any toll charges incurred while transporting your possessions. If you are traveling a long distance to your new home, this may not be an option. Conversely, with short distance moves, maybe it is worth calling in this favor.

Tips For Choosing a Mover

Finding the right mover is just as difficult, if not more, than finding the right contractors to work with you on building your new home. Because this process is so time-consuming, it is important enough for you to make sure lead-time is accounted for in your timeline. The last thing you want to experience is delays associated with moving into your new home, especially after all the red tape you have just gone through during the final inspection.

Here is a list of points to follow when choosing the right mover on moving day:

1. Get a written estimate from between three and five movers. You will be approaching this phase of the process exactly the way you did during the selection process for contractors. This estimate should be based on the actual inspection of the possessions you will be moving from point A to point B. You also want to know if the mover charges per pound or per hour. Inquire about charges that may incur if you are planning to move on a weekend.

2. Look for registrations and insurance. These requirements may vary, particularly if you are moving your possessions from one state to another. The U.S. Department of Transportation (DOT) requires a Federal Motor Carrier Safety Administration (FMCSA) number.

To learn more about this requirement, visit **www.fmcsa.dot. gov.** Learn about what is and what is not covered by the insurance they have in place.

3. Research complaints about the mover. You can obtain this information from the Better Business Bureau (BBB), or other consumer protection agencies listed in your phone book. Conducting a search using your favorite search engine may yield some useful information about the mover's history, too.

4. Put your own priorities first. Do not sacrifice the quality of service by choosing the cheapest mover. Just like when making the choice between contracting bids, you will get what you pay for during this process as well.

Are There Red Flags to Look Out For?

There are warnings when a company is not the one you want moving your goods. This is true in any situation, so it cannot be overlooked in this one. As tempting as it may be to ignore red flags and be trusting, you cannot afford to be naive. Your money is being invested and you cannot go over budget when you are so close to the finish line. Here are some situations that should alarm you and cause you to consider rejecting the contract:

• The moving company estimates the cost of the move over the phone and plans no on-site visit. It will save them time and make them money. If you have to ask them to come out to estimate, then you are dealing with the wrong people. Look elsewhere.

• Large cash deposits or up-front payment is demanded. This is not ethical, nor is it required in order to do business. If you experienced this situation, it is a definite red flag to move on.

• You are not provided with a copy of the booklet titled, "Your Rights and Responsibilities When You Move." Movers are required by federal regulations to provide this information, and failure to do so is a red flag.

- Lack of information on their Web site. If they claim to be local movers but they have an out-of-state address listed, that is a red flag. In addition, there is cause for concern if they do not list their proof of insurance or the registration numbers.

- They say their insurance covers "everything." This is false and another red flag to move on.

What is My Recourse if I Experience a Red-Flag Mover?

If you have experienced any of the issues listed above, it is important to take action. Not only will this resolve your issue or put you closer to a resolution, it will protect other consumers from going through the same experiences. Here are some contacts you should make in the event of experiencing a red-flag mover:

- File a complaint directly with FMCSA through their Web site, **www.fmcsa.dot.gov**, or by calling 1-800-832-5660.

- Call the Department of Transportation Authority: 1-800-DOT-SAFT.

Other Hints For a Smoother Move

- Keep phone numbers handy for the movers or anyone else helping you. Cell phone numbers are best, but if they are not available, direct lines will do.

- Have cash or traveler's checks on hand for anything unexpected that might come up. Pay for meals and fuel.

- Important documents related to the sale of your previous home, as well as the certificate of occupancy for your new home, should be kept with you at all times.

- Bring a copy of your current phone book to keep on hand in case you need to reach utilities companies. It is best to be prepared during an emergency situation.

- An overnight box should be packed in case there are delays with the movers or any other aspect of moving day. This box should contain medications, vital documents (such as birth certificates and insurance policies), toiletries, towels, changes of clothes, and non-perishable food items.

- Transport your valuables yourself if you can. Normally, this would include jewelry, photo albums, computer data, and paperwork containing financial information. Use your best judgment and bring along anything you have second thoughts about.

Money-Saving Tip:
Just like planning all the necessary phases of your home-building project, preparation for the move should be just as meticulously considered. Like being an owner-builder, you will save money by taking the time to plan out every aspect of your move well in advance.

15.4: Setting up a Maintenance Plan

Prepare a year-long maintenance schedule based on where you live. This may seem like a hassle, but you have learned from the building process how crucial good planning is. After completing a huge undertaking, the last thing you want to deal with is being unprepared. Some questions you should ask yourself include:

1. How often does the furnace (or other heating unit) need to be serviced?

2. What needs to be done to keep my appliances in optimal condition?

3. If the house has a chimney, how often does it need to be cleaned?

Gather all the service manuals and guides you have been collecting and filing each time an installation occurred during the build. These will help construct the basic skeleton of your home maintenance plan. Then make a checklist outlining other significant areas of the home. This list should include:

- **Roof:** Over time, leaks could occur around the chimney or any skylights. In addition, you have to check for ice forming under the shingles along the gutter line.

- **Attic:** If there is not proper ventilation, problems with mold and mildew can occur. Should you be unsure of what to look for, check with a specialist to ensure you are not passing over problem areas.

- **Basement:** Dampness may occur, so check the walls and install a dehumidifier if necessary. You should also check for any leaks, as well as assess the potential need for a sumo pump[1].

- **Exterior paint or siding:** If you painted, check for cracks or holes. With siding, it may be necessary to replace caulking to prevent basement flooding and other issues.

- **Gutters:** Be sure all gutters drain away from the house and that they are clear of leaves and debris. Otherwise, you risk the chance of ice buildup.

---

[1]Sumo pumps are automatically triggered when water reaches a designated level. They ofen are connected to drain pipes, out basement windows, or through outlets drilled into the wall.

- **Filters:** Change them as often as recommended by the manufacturer. If there are no maintenance guidelines available, seek the advice from the installation contractor or research this information on the Internet.

- **Doors and windows:** Inspect all doors and windows for leaks to ensure better heating and cooling efficiency. These leaks could potentially cause water damage.

- **Detectors and Extinguishers:** Check all safety equipment regularly to ensure they are all in working order. Replace batteries in smoke alarms whenever clocks are set ahead or forward as a good reminder.

Notes From the Field

Some owner-builders who decided to hire a general contractor have reported they provided the homeowner with a home maintenance plan with the final bill. If the general contractor you hired does not offer this to you, inquire about help with setting one up.

You may also wish to organize your maintenance plan by season. This is useful for those who keep daily planners, home repair binders, or home task organization wall calendars consisting of the various chores they would like to accomplish on a seasonal basis. These are some examples of what could be on seasonal checklists:

- **Spring:** Check the exterior of the home for any obvious defects, make repairs on screens, and inspect weatherproofing materials. Interior inspections include changing filters, cleaning the chimney, checking for leaks in pipes and faucets, looking for dampness in the basement, inspecting for pests, and maintaining the refrigerator.

- **Fall:** Inspect weatherproofing materials, inspect the exterior of the home for any visible damage, remove and maintain screens, install storm windows and doors, clean gutters, and check chimney for damage. Interior inspections include checking the fireplace flue, inspecting insulation in attic, maintaining humidifiers, maintaining hot water heater, and maintaining the refrigerator.

For specialized checklists customized to your home, specific areas, and seasons, visit the Internet. There are hundreds of Web sites available offering customized lists, tips, and advice. These are printable, so creating a home maintenance binder is quick and easy. This is going to become a well-utilized and significant part of the operation of your home, so it will be necessary to accomplish this task.

Money-Saving Tip:

To ensure your appliances are running efficiently and not running up your energy bill, keep them clean. Lightly vacuum behind the refrigerator, clear dust from vents, and make sure dryers are free from dust and debris. By staying on top of maintenance likes this, you will see an improvement in your energy bills. This housekeeping practice will also increase the longevity of your appliances, which is also a money saver.

CHAPTER 16
Spending Now or Later

Contracts have been discussed throughout this book, including do-it-yourself contracts, contracts obtained off the rack at a stationary store, and free or purchased downloaded contracts.

Every state has different requirements and legal terms. While this book is designed to help the home-builder save thousands, there are some expenses you should not avoid. Whether you are acting as the general contractor or hiring one, there are laws, legal consequences, and liabilities. Just as you need insurance to cover all possibilities, consider using an attorney. As a home builder acting in the capacity of general contractor, you have obligations different from those of someone who holds a license. What kind of protection are you building in for unseen hazards? Who will be responsible for resolving issues? Who will provide the insurance for subcontractors? Will you have an arbitration or mediation clause, and do you know the difference? Do you know the advantages or disadvantages? When the general contractor presents you with a contract, do you understand everything it says and how it will affect you?

CASE STUDY: ANDREA GOLDMAN, ESQ.

Attorney

Newtonville, Massachusetts

Andrea Goldman is an attorney in Newtonville, Massachusetts, who specializes in dispute resolution between home builders and contractors. She said most disputes between contractors and home builders arise from failure to clearly define roles and responsibilities, plus a lack of communication. A home builder who is acting as his or her own general contractor is clearly at risk if he or she is not aware of legal accountability, as is anyone who signs a binding document without the sanction of an attorney.

One particular factor in negotiation after the fact is the lack of understanding by one or both sides when agreeing to binding arbitration. In a dispute, each party presents their side and the arbitrator makes the decision. While one would expect the resolution to be fair, what if the decision was biased? Depending on your state, the only recourse may be if the rendering circumvented the law and was therefore illegal. Many people agree to an arbitration clause without comprehending its finality and short-falls. Mediation, on the other hand, is where the parties discuss their differences with an independent mediator who helps both sides resolve the disagreement, and nothing is binding without the consent of both parties.

Contracts are important when it comes to timelines, among other issues. If the project is given six months and goes eight, will there be penalties, and how will they be enforced? What if you fire the contractor — will they keep your deposit? If so, how much will be paid? What if, like in an earlier case study, a surveyor makes a mistake and it is not detected until much later in the process? It is much more expensive to retain a lawyer after the fact rather than to work with one from the start. Furthermore, finding an attorney during the planning stage gives you time to research; waiting for a pending lawsuit puts you under pressure, and you may well select someone who is available rather than is able.

These are just some of the issues keeping Goldman in a busy practice. While many people see lawyers as unnecessary, home builders soon develop a healthy respect for their capabilities in preventing litigation with solid contracts. As the adage goes, you can pay a few hundred dollars for a contract or $10,000 in attorney fees during litigation.

For more in-depth discussions on the intricacies of building litigation and prevention, see Goldman's Web sites and blog: **www.andreagoldmanlaw.com**, **www. buildingconfidence-llc.com**, and **www.andreagoldmanlaw.blogspot.com**

CASE STUDY: S. G. WARDONE

Fishkill, New York

In the matter of Sara Wardone, the story is an example of the need for competent legal advice and common sense before agreeing to anything.

Gail tells of her lifelong dream to have a large vegetable garden. Come spring of that first year, she learned the builder had dug up all her fertile soil during the excavation, sold it to others, and back-filled the acreage with garbage, dirt, and rocks. By court order, the contractor graded the property, as well as the properties he sold to others in the neighborhood, and left them with a minimal 3 inches of planting soil. That was the tail end of the story.

Had the Wardones done sufficient due diligence before signing with the builder, they would have found out though their contractor could point to prior homes he built, none of the owners were happy with his work and were in the process of seeking legal action against him for a multitude of complaints.

We could probably write a book about this contractor and his unethical practices, but let us just stick with the highlights. When the house was being framed, the building was not facing the same direction the owner and builder agreed to. Upon completion of the framing, the Wardones discovered the builder had altered the plans such that the staircase leading upstairs was moved, and the formal dining room lost space to the kitchen, which was expanded. It may not seem significant but the domino effect meant there was now a gap between the wall and staircase. The builder insisted on $250 to finish it off and make it a small closet. By moving that wall to make the kitchen bigger, the exhaust vent, which originally would have been hidden in the joists, now stuck out. The builder refused to enclose the area, and the owner had to do it himself. And because the kitchen was expanded, the cabinets that had been ordered left a large gap of about a foot at the other end of the wall.

The build cost more than $100,000 over the original budget. The Wardones had an ineffective attorney and a builder who should not have a license. This story should not scare you away from the process, but should warn you of the dangers inherent to believing you are astute enough not to need an experienced attorney.

CONCLUSION

Now that you have completed this book, there should be no question about what is involved in becoming an owner-builder for your own new home. It is surely a challenging process, but it is not impossible, and you can accomplish this without losing your mind in the process. With any challenge comes hard work, so remain confident in your abilities, remain realistic, and you will be fine.

Before finalizing paperwork, have a meeting with the contractors to work out all the bugs. As with any other large-scale project, it is unlikely this project was completed without errors. It is necessary to review everything with your contractors point-by-point to ensure all the problems have been successfully resolved. Some of these can be ironed out during the final weeks of construction, while others will have to be discovered after the process has been completed. This final meeting should occur at least one week prior to the closing. If this is not feasible, schedule this meeting as close to that time frame as possible.

This meeting also serves to ensure that all the provisions of the contract created between you and your contractors have been met and performed. A good contract is one that will have addressed the issues of problems found post-construction and responsibility for addressing them within a specified time period. If there are items in the contract that have not been completed due to weather or other unforeseen reasons, ask the contractor for a schedule of expected completion. Follow these tips for finalizing things with the contractor in an amicable manner:

1. **Speak to the contractor as calmly as possible:** Keep a level head at all times no matter how upset you are or what response you hear.

2. **Suggest mediation:** If terms cannot be agreed upon amicably, request mediation in which a mediator sits in during a discussion. This will allow for a productive conversation that is likely to produce finalized results.

3. **Secure an attorney:** This is the last resort, once all other options to settle things in a civil manner have been exhausted. Sometimes mentioning your desire to obtain an attorney is enough to promote a resolution.

Ask for any explanations needed for operation of heating and cooling units, where important items are located such as the water softener, shut-off valves, and also appliances you are unsure of. Be sure all the warranties have been provided and explained to you. If you have a septic tank, find out where it is, where the leech field is, and where the connections are. You will have to know where the well is and how to access it if needed.

Notes From the Field

Unless you are an attorney, it is best you do not draw up your own contracts. A good contract can avoid post-construction problems and the need for litigation to get issues resolved. Lay individuals do not know the nuances of the law.

The year following the build is when you must keep a keen eye on the house for any structural flaws that could potentially develop into something more serious. During this year, do not be afraid to contact the builder to resolve these problems. If these damages are on portions of the home you built yourself, find out how to repair the trouble through advice from contractors or guides. The ultimate goal is to resolve issues as hassle-free and cost-effectively as possible.

Now that you have accomplished all your goals, it is time to share your knowledge. Share your good fortune with those who hope to achieve the same. When you were going through the home-building process, you may have been missing one key person: a role model or mentor. If you did not have someone like that and found yourself in need of one, now is your opportunity to play that role for someone else.

Showing your appreciation for the contractors, subcontractors, and tradesmen's work adds the finishing touch to the home-building process. You can approach this in a couple different ways: Send thank-you notes to each of them expressing your gratitude and pointing out the good work they accomplished, and request business cards from them so you can promote their business to other owner-builders.

The points outlined in this book are intended as a guide to get you on your way. Use your instincts, read up on the subject as much as possible, and use all available resources. Remember to keep your options open, especially if you feel you are in over your head. With the proper planning, support, and guidance, you can accomplish the goal of building your own home and saving thousands of dollars.

When everything is complete, consider creating a five-year plan dedicated to your home. This plan should contain many of the same building blocks that your owner-builder plan was built on. Here are some examples for the contents of a good five-year plan:

1. Determine what you would like to add on to the home in terms of rooms, amenities, and cosmetic features that the budget did not allow for in the original build.

2. Draw up some landscaping plans that you would like to accomplish over the next five years that the original budget did not allow for, or that your original timeline did not permit.

3. Set up a time frame for when these building and landscaping projects should occur. Piece the projects out so they are spread evenly over the course of the next five years, giving special attention to the seasons and when would be the best time of the year to accomplish these goals.

4. Make a list of materials and equipment necessary to accomplish each task on your to-do list. If you can borrow any of these items, or if any of these items are left over from the original build, note it on your list.

5. Create a budget that will allow for these projects to be completed gradually over the designated time period. If your budget does not allow for all of these projects to be completed over the span, create a plan that extends for a longer period of time.

With a plan like this in place, your role as an owner-builder does not necessarily have to end when your new home is move-in ready. Creating this plan will also allow you to incorporate all the elements that you were not able to include during the original build. This is going to alleviate a lot of potential for regrets you may be worried about lingering on the horizon.

Your building project is going to be very personal to you. Utilizing the tools and tips outlined in this book will help you not only be successful in your journey toward being an owner-builder, but it will also give you the opportunity to have exactly what you want and what you need. You will be doing what is right for you and those who are touching your life during every step of the way.

Sample of a Subcontractor Suggested Interview Questions:

1. What do you consider to be a good job?

2. Why do you think you are better for this job than other subs?

3. What do you need to do in order to produce a bid?

4. What kinds of costs are typical for a job like mine?

5. What kind of cost-saving tips do you have?

6. What kind of tools or materials do you own that would save money?

7. Are there other subs you like working with or that you normally work with?

8. What do I need to do to make this job easier for you?

9. What kind of schedule do you work with?

10. How big of a crew do you work with?

11. How much experience do the people you work with have?

12. What kind of time frame do you think this job will have?

13. Do you guarantee your price and schedule?

14. Are you insured?

15. Our projected start date is _____; does this work for you?

16. What other suggestions can you offer for this job?

17. What is the best way to reach you (cell phone, work phone, e-mail, fax, mail) for sending plans?

Sample of a General Contractor Interview Questions:

1. Can you provide a list of references? (Ask for between three and five).

2. Have you worked on any projects similar to mine?

3. Do you have a project portfolio you can show me?

4. Do you have a crew or do you work from a stable of subcontractors?

5. Will you be willing to work with subcontractors I have chosen?

6. Will you be working on this project exclusively?

7. If not, how many job sites are you normally on in a given week?

8. How much time can you dedicate to working on-site daily?

9. Will everyone on-site carry the proper licenses and insurance?

10. How do you work with customers to save time and money?

APPENDIX B
Checklists

SAMPLE OF A FINAL INSPECTION PUNCH LIST

(Add or remove items as necessary for your project.)

INITIALS	FINAL SIGNOFF FROM ALL INSPECTORS
	Building Inspector
	Electrical Inspector
	HVAC Inspector
	Plumbing Inspector
	Termite Inspection
	Other:
NA/TBC	**Exterior**
	All concrete forms have been removed.
	Grade slopes away from house 6 inches vertically for every 10 feet horizontally.
	Final grade smooth and back fill settled.

NA/TBC	EXTERIOR
	All wood products removed from soil around house (termite protection).
	Loose concrete cleaned up and hauled off.
	Hose bibs work.
	Brick and other wall surfaces cleaned.
	Gas and electrical meters are connected and sealed.
	All penetrations through walls caulked and sealed.
	Gutters and downspouts properly installed.
	All exterior trim properly installed.
	Shingles correctly nailed.
	Roof flashings properly installed.
	Roof vent collars caulked or sealed.
	All fascia correctly installed.
	All soffits correctly vented and installed.
	Windows securely installed and caulked including top; windows operate correctly.
	Screens installed.
	Siding properly nailed without bulges.
	Trim properly nailed and caulked.
	Deck properly installed and nailed.
	Steps and railings secured.
	Exterior locks working.
	Exterior doors were properly hung; they properly close and properly lock.
	Door weather stripping and thresholds installed and adjusted.
	Garage doors properly installed and operating.
	Site is clean.
	No missing insulation in floor or ceiling joists.
	If elevated structure, is drainage and ventilation adequate?

NA/TBC	INTERIOR
	All doors are plumb, open freely, and do not close by themselves.
	All bypass and bi-fold doors are plumb and properly operate in their track.
	All doors clear the carpet.
	All doors undercut for air circulation.
	Door stops installed.
	Base joints tightly fit and are caulked.
	All shelving properly installed.
	Closet rods installed.
	All drywall joints and fasteners properly floated.
	All trim and moldings in place and properly installed and caulked.
	All surfaces requiring paint are properly painted with adequate covering.
	No gouges in wall from installation or finish work.
	Counter tops correctly installed.
	Insulation properly installed in attic.
	Attic access properly installed and insulated.
	Outlets and switches installed — no gaps in cover plates.
	Lights and outlets properly operate.
	All fans work.
	Dishwasher works.
	Oven and stove properly operate.
	All light fixtures have bulbs in them and work (bulbs are not covered by the warranty).
	All electrical trim in place.
	Proper disconnects in place and operational.
	All electrical circuits labeled.
	All smoke alarms function.
	All HVAC registers in place.

NA/TBC	INTERIOR
	Clean filter in A/C and furnace.
	Thermostat controls installed and operational.
	Furnace turned on in the winter.
	A/C on in the summer and certified by contractor as able to keep the house 20 degrees below outside temperature.
	Ductwork properly installed, insulated and sealed.
	A/C fans operate properly with acceptable noise levels and good distribution.
	Proper plumbing shutoff valves in place and operational.
	Tubs and sinks properly installed and caulked.
	Hot and cold water in all sinks turn on.
	No leaks in plumbing fixtures.
	Toilets properly are installed and flush properly.
	Floor drains have water in traps.
	Plumbing access panels installed.
	Bath hardware tightly fitted to the wall.
	Mirrors securely installed.
	Shower doors installed and caulked (where applicable).
	Appliances properly installed and working.
	Cabinet doors close.
	Cabinet doors and drawers operate properly — open all cabinets and eliminate squeaks.
	Cabinet trim properly installed.
	Counter tops installed and properly caulked.
	Linen closet shelving correctly installed.
	All flooring installed properly and without damage.
	Rails and balusters securely fastened.
	All windows open and close without binding.

NA/TBC	INTERIOR
	All windows lock.
	Floors vacuumed clean.
	Handrails and/or guardrails are installed and properly secured on all stairs.
	All parts of the house clean and free from construction debris.
	All excess material neatly stacked in the garage.
	Review the contracts and check all change orders to make sure they are completed.
	Manuals for appliances and systems ready to give to owner.

APPENDIX C
Worksheets

Sample of a Budget Worksheet

EXTERIOR FINISH:	Bid:	Changes:
Roofing		
Siding		
Gutters		
Decking/porches		
FLOORING:		
Ceramic tile		
Carpeting		
Hardwood		
TOTAL:		
Other costs:		
Overall total:		
Appraisal:		
Equity:		

Sample of a Lien Waiver

WAIVER OF LIEN
Material or Labor

Home Builders
Project: _____
Job Site & Number: _____
Subcontractor: _____

Contract Price:	$_____
Authorized Extras	$_____
Previous Payments	$_____
Requested Payments	$_____
Balance Due	$_____

For value received the undersigned hereby waives and releases all statutory or equitable liens or right to liens as to the above described premises to the extent of the total payments received as shown above for work, material, and services furnished.

The undersigned hereby waives any and all legal or equitable defenses that might defeat this waiver and waives any right to liens for work, material, or services not covered by the contract price.

The undersigned warrants that the work, material, or services furnished by the undersigned are equal in value to the total of all payments received as shown above.

IN WITNESS WHEREOF, the undersigned executed this agreement on the _____ day of _____, 20___.

X	X
SUBCONTRACTOR	WITNESS
X	X
Home Builder	WITNESS

Revised (01/07) - www.thebuildersurvivalkit.com/documents/Sample-Lien-Waiver.doc

Sample of a Monthly Planning Worksheet

MONTH: _____   **PAGE#:** _____

Week at a Glance:	ACTIVITIES:

Sample of a Weekly Planning Worksheet

TASK:	Week of:	Week of:	Week of:	Week of:	Week of:
1.					
2.					
3.					
4.					
5.					
6.					
7.					
8.					
9.					
10.					
11.					
12.					
13.					
14.					
15.					

Sample of a Weekly Planning Worksheet

TASK:	Week of:	Week of:	Week of:	Week of:	Week of:
1.					
2.					
3.					
4.					
5.					
6.					
7.					
8.					
9.					
10.					
11.					
12.					
13.					
14.					
15.					

BIBLIOGRAPHY

The Owner-Builder Book: How You Can Save More Than $100,000 in the Construction of Your Custom Home, by Mark A. Smith with Elaine M. Smith, 2007, The Consensus Group Inc.

Building Your Own Home For Dummies: The Complete Custom Home Guide – from foundation to financing, by Kevin Daum, Janice Brewster, Peter Economy, 2005, Wiley Publishing, Inc.

Build Your Own Home. A Concise Guide to Successfully Subcontracting and Building Your Own New Home, by Michael A. Pompeii, 2006, Michael A. Pompeii, P.E.

Tips & Traps When Building Your Home, by Robert Irwin, 2001, McGraw-Hill

How to Plan, Contract, and Build Your Own Home, Fourth Edition, by Richard M. Scutella and David Heberle, 2005, McGraw-Hill.

InfoForBuilding.com: Making it Easier to Build Your Home, 1999-2008 Construction Documents Company, Lien Waiver **www.infoforbuilding.com/ Lien_waiver_N.html**.

Frank B. Norris & Company: The Builder Survival Kit, Sample Lien Waiver. **www.thebuildersurvivalkit.com/documents/Sample-Lien-Waiver.doc**

The Road Home: Building a Safer, Stronger, Smarter Louisiana, Punch List, Copyright 2008, *The Road Home Program*, Posted on: 1/18/2007.

http://dc.road2la.org/en/family_home/home/design_construction/ Getting+Started/Construction+Process/Punch+List/

BIOGRAPHY

Corie Richter has been a freelancer writer since 1998 both for print and online publication. She was first attracted to green building when living in an adobe home located in New Mexico. The community presented the opportunity to participate in construction of adobe, forged-earth, and baled hay projects as part of an eco-healthy environmental effort. The author realized that with a basic knowledge, anyone could get started doing their own work; all it took was fortitude and knowing when to ask for assistance. The home-building process has been a passion since then, and she looks forward to sharing her knowledge and experience with others, as well as building another home.

Corie is the author of *God Has a Sense of Humor*, a comedic book reflecting on the foibles of life, and is noted for her many articles and talks addressing health care issues, nutrition, and aging. In addition she has plied her craft to educational study guides, test-question writing, and contributing to academic publications. The Army veteran enjoys the company of her companion Drummer, who was adopted from a Virginia animal shelter where he was brought after an abusive owner forced the young dog into fighting, then abandoned on a roadside. The author holds strong convictions regarding man's stewardship of the Earth and its creatures.

INDEX

A

Algae, 243

B

Backfilling, 152-153, 8
Bathroom Vanities, 201, 9
Bed Molding, 186
Brick Veneer Siding, 175
Building Lot, 83, 86, 95, 6

C

Case Molding, 186
Catch Basins, 154
Caulking Irons, 178
Cedar Shingle Siding, 175
Ceramic Tile, 277, 37, 200, 126
Cloat Head nail, 161
Construction Loans, 75-77, 79, 6

Convection, 226
Cooling Systems, 127, 131, 160, 7
Cornice Lighting, 218
Corrugated Fastener, 161
Crown Molding, 186

D

Decision Matrix, 90
Decks, 106, 128, 171, 173, 8
Demolition, 96-97
Depressions, 237
Direct-vent Gas Fireplaces, 140
Doing Business As (DBA) license, 67
Dome Home Construction Loans, 76
Domino effect, 261, 63
Down Lights, 219
Drainage, 272, 42, 178, 232, 236,

242, 89-90, 128, 148-149, 152-154, 166, 172, 8

Drainage Gravel, 154

E

Electrical Thermal Storage Units, 126

Environmental Protection Agency, 132, 151

Excavation, 238, 242, 261, 90, 145-146, 154, 8

F

Fiberglass shingles, 165

Fireplaces, 122, 136-140, 7

Flashing, 139, 166, 170

Flashing, 139, 166, 170

Footers, 146, 8

Foundation, 281, 36-37, 42, 181, 195, 231, 236, 58, 84, 94, 98-100, 109, 116, 146-155, 166

Foundation Plan, 109

French door, 225

Frost Line, 147, 151, 172

G

Galvanized, 170

Garbage Disposal, 229

General Contractor, 268, 17-19, 21-22, 25, 32-36, 256, 259-260, 44, 49, 54, 59, 65, 70, 73, 76-77, 109, 5, 11

Gypsum Board, 182

H

Hazardous waste, 96

Heat Pump Units, 125

Heat Transfer Sites, 181

Heating Systems, 124-126, 7

Hire-and-fire, 32

Humidifiers, 257, 132-134, 7

I

Insect infestation, 149

Insulation, 272-273, 39-40, 181-182, 191, 204, 257, 159-160, 164, 168

Interior Priming, 223, 9

L

Laminate, 188, 196-197, 200

Large Dumpster, 70

Light-duty Truck, 72, 171

Log cabins tend., 165

Low Bidder, 113, 150

M

Masonry Fireplaces, 138

Measure twice and cut once, 63

Metal Shingles, 165

N

National Ground Water Association, 151

O

Owner-Builder Construction Loans, 76, 79

P

Pilot Holes, 210, 170
Plumb bob, 179
Plumbing, 271, 274, 27, 29, 31, 39, 177-178, 180, 191-192, 202, 204-205, 231-232, 58, 95, 150-152, 8-9
Portable Electric Generator, 70
Post and pier, 148
Post-construction budget plan, 249
Pre-hung windows, 169
Prime real estate, 97
Pump station, 101

R

R-value, 181-182, 164
Radiant Floor Heating Systems, 126
Raw land, 91
Recessed Lighting, 194, 219

S

Seepage, 166
Sheathing, 159-160
Shower Nozzles, 192
Single Close Construction Loans, 75
Slabs, 239, 146-147, 8

Solar Units, 125
Square, 179, 187, 208-209, 216, 225, 71, 78, 126, 144, 161, 172
Stairways, 206-207, 210, 9
Standing Trim, 186
Steel rebar, 148, 156
Stone Siding, 176
Stucco Siding, 176
Subcontractor bidding process, 79

T

T-blocks, 40
tax identification number, 66-67
Top-mount freezer refrigerator, 225
Tubular grates, 141

U

Underground Plumbing, 150-152, 8
Uniform Building Code, 108
Up Lights, 219

V

Valance lighting, 217
Ventilation Systems, 160

W

Wire Mesh, 238
Wood clapboard siding, 176
Wood Clapboard Siding, 176